A WORLD
WITHOUT
BEES

Alison Benjamin is editor of *Society Guardian* and writes on social affairs and environmental issues for the newspaper.

Alison took up beekeeping four years ago with her partner, Brian McCallum who has just become a geography teacher.

They keep a number of hives in London, are the authors of *Keeping Bees and Making Honey* and have a blog at www.aworldwithoutbees.com

Praise for *A World Without Bees*:

"A timely and important exploration of the crucial role bees play in all our lives and the deadly threats they are facing."

– Vi

A WORLD WITHOUT BEES

ALISON BENJAMIN
AND BRIAN McCALLUM

guardianbooks

Published by Guardian Books 2009

2 4 6 8 10 9 7 5 3 1

First published in Great Britain in 2008 by Guardian Books

Guardian Books
Kings Place, 90 York Way
London N1 9GU
guardianbooks.co.uk

A CIP catalogue record for this book is available from the British Library

ISBN: 978-0-85265-131-5

Typeset by seagulls.net
Printed and bound in Great Britain by CPI Bookmarque, Croydon, Surrey

CONTENTS

INTRODUCTION

WHY A BOOK ABOUT A WORLD WITHOUT BEES?

For as far as the eye can see, everything is pale pink. The valley that stretches across central California for the best part of 400 miles (650km) is blanketed in salmon-coloured orchards.

Welcome to almond country. The trees – all 60 million of them – are heavy with blossom. Other than a constant stream of cars and trucks along Route 5, and the fast-food joints that hug it, there is little to see in this flat landscape beyond row upon row of flowering trees.

When we told people we were coming here, the usual response was: "Wow! That's going to be beautiful." They were right about the "wow" factor: almond-growing on this scale is mind-boggling. But where is the beauty in such a regimented landscape?

The trees are planted in symmetrical rows, at regular intervals, so many feet apart. Early-blooming and late-blooming varieties are laid out in separate blocks, in uniform, repetitive patterns.

This standardised, industrial-scale method of producing a single crop, known as monoculture, has become the hall-mark of modern agriculture. Adopted across the globe, it has led to substantial increases in food supplies. Yet few crops can match the inexorable rise of the Californian almond, which is now the US's most valuable horticultural export. In 2006, more than $1.9 billion-worth (£1.4 billion) of almonds were sent from California to the global marketplace – double the revenue from its Napa Valley wine exports. In fact, 80% of the world's almonds now come from the golden state.

This was not the case just 30 years ago, when an acre of almond orchard produced around 500lb (225kg) of nuts. Today, average yields six times that are not unusual. But it is not just better management or new varieties that explain these record-breaking harvests. If you look closely at the blossom-laden branches, you will see the reason for the explosion of fruit. And if you listen, you will hear the unmis-takable buzz that accompanies the diligent work involved. For each flower has on it a honeybee. She is drinking its sweet nectar. As she crawls around to find the perfect suck-ing position, her furry body is dusted with beads of pollen that will be transferred from blossom to blossom, fertilising the flowers as she makes her rounds. By late August, ripe oval-shaped nuts will show where she has passed.

Apis mellifera, or the western honeybee as she is more commonly known, has been revered for thousands of years for her ability to make a sweet substance that delights the human palate. The earliest record of humans' use of honey is a cave painting in Valencia, Spain, depicting a man climbing a cliff to rob a swarm of wild bees. It dates back at least 10,000 years, to just after the last ice age, and the love affair has continued ever since.

Most people, however, know little about the honeybee's importance as nature's master pollinator. The majority of flowering plants need animals to pollinate them, and the honeybee is perfectly engineered to perform the task, with a body designed to trap pollen and a work ethic that leaves no petal unturned. Without the honeybee, the vitality and colour of the planet would be lost.

But it's not just pretty flowers that we need to thank honeybees for. Approximately one-third of the average diet – or, to put it another way, one in three mouthfuls that the average person eats – has been pollinated by them, from nuts, soya beans, onions, carrots, broccoli and sunflowers to apples, oranges, blueberries, cranberries, strawberries, melons, avocados and peaches. Alfalfa, the clover-like plant widely grown as cattle feed, is also dependent on the honeybee, as is cotton. In all, some 90 commercial crops worldwide owe their continued existence to the honeybee. That makes honeybee pollination worth more than an estimated $60 billion (£43 billion) a year, of which some $15 billion (£11 billion) is in the US alone, according to a study by Cornell University.

Pollination is big business, and nowhere more so than across the 600,000 acres (240,000 hectares) of Californian almond trees. Each February, they play host to around 1.2 million honeybee colonies. Each acre houses two hives: that's some 80,000 bees per acre, or more than 40 billion bees in total, making it the largest pollination in history.

We'd been told it was an amazing spectacle. But unlike the sight of tens of thousands of migratory birds flying south for the winter, the arrival of billions of honeybees in the warm climes of Central Valley is not a natural phenomenon. They are guided neither by the position of the sun, nor by the Earth's magnetic fields. Instead they are driven thousands of miles on the backs of huge trucks from the far corners of the United States, 500 hives at a time, stacked four high.

Half of all the 2.4 million honeybee colonies in the US make this annual cross-country trek from as far afield as Massachusetts in the east and Florida in the south. They are now joined in the Central Valley orchards by honeybees flown in from Australia.

And it doesn't end there. California is just the first port of call on many of these bees' five-month tour of America, which takes in more than 3.5 million acres (1.4 million hectares) of orchards and fields. After three weeks feeding on almond nectar, many will be back on the trucks heading south to the citrus plantations of Florida, then north for apples and cherries, and as far east as Maine for the blueberries.

This intensive, migratory beekeeping is a far cry from the hobby that the two of us pursue in our small back garden in south London. The only move for our bees was from the apiary where we collected them to the spot by the wall where their hive has sat for a couple of years. From this sheltered location, they happily forage from spring right through to the end of autumn for nectar and pollen among the parks, gardens, railways sidings and tree-lined roads that dot the Battersea landscape. In the process they make enough honey to keep us and them well fed throughout the year.

There is something magical about watching your bees return home after a hard day's foraging on a balmy summer's evening. For many urban apiarists who work all day in an office, they are an antidote to the stresses of city life. Creating a rural idyll in a corner of a housing estate was our small way of trying to reconnect with nature. It fulfilled something we knew was missing from our lives, a feeling that we couldn't quite put our finger on, but is now being termed "nature-deficit disorder".

We had also heard about the vital role that honeybees play in the ecosystem, and the threat they faced from the same combination of factors that afflicts much of our wildlife – urban development, loss of biodiversity and destruction of habitat. So by giving bees a home in the city it felt as if we were doing our bit for the environment.

There is nothing vaguely eco-friendly, however, about trucking billions of bees thousands of miles every year. The contrast between our "back to nature" vision and the harsh reality of commercial beekeeping is unfathomable.

What is happening in California is nothing short of the industrialisation of pollination. And, like any industry, it is driven by profit. Commercial apiarists can clear more than $100,000 (£72,000) for just three weeks' work, and the farmers' income rises as yields increase.

Joe Traynor is a bee broker. For six weeks every year, his company, Scientific Ag, matchmakes migratory apiarists with Californian almond farmers in need of bees. It is testimony to the scale of the almond industry that it has spawned a whole new career for Traynor and other former beekeepers.

But now the almonds, and other crops that rely on bees for pollination, are menaced by a mysterious illness that has led to the disappearance of billions of honeybees around the world and is fuelling fears of an environmental crisis bigger than climate change.

Albert Einstein is reputed to have said: "If the bee disappeared off the surface of the globe, then man would only have four years of life left. No more pollination, no more plants, no more animals, no more man." In truth, it is more likely to have been French beekeepers who put these words posthumously into Einstein's mouth a few years ago during a battle to get a pesticide banned from their country. Whoever said it, however, the apocalyptic sentiment chimes with the view that bees are the "canary in the coal mine", a guide to the health of the planet, and that their predicament is a warning to us all.

For two years running, a third or more of all honeybees in the States have mysteriously died – around 800,000 colonies in 2007 and one million in 2008. Some commercial beekeepers have reported losses of up to 90% since the end of 2006. The disappearances, which have baffled researchers and academics, are not limited to the US. Large numbers of colonies have also been wiped out in parts of Canada, Europe, Asia and South America. In Croatia, it was reported that five million bees disappeared in less than 48 hours.

Bees have a sophisticated navigation system that uses the sun and landmarks as points of reference. It allows them to travel up to three miles (5km) from the hive in search of food without losing their way back home. They are able to direct other bees to food supplies through a remarkable form of communication called the "waggle dance". But in a hive suffering from this strange plague the adult bees do not return home, leaving their queen, eggs and larvae to starve to death. Moreover, young nurse bees, whose job it is to stay in the hive and care for the newborn bees while the adults are out searching for food, desert their posts and fly away. Such a dereliction of duty is unheard of unless the bee is diseased and leaves the hive to avoid infecting others.

When news of the vanishing bees – a phenomenon soon dubbed colony collapse disorder (CCD) – started to filter through in newspaper reports at the beginning of 2007, some of the more fanciful theories for the disappearances ranged from cellphones messing up the bees' navigation system to an elaborate al-Qaida plot to wreck US agriculture.

Although no one knew for sure what was causing the bees to perish, it spurred the launch of a global investigation. More credible suspects included exposure to genetically modified crops, pesticide poisoning, invasive parasites, malnutrition, and the stress of being moved long distances. Entomologists were convinced that the culprit was either a new virus, a virus that had mutated into a more virulent strain, or a virus that had combined forces with another pathogen, such as a fungus, to create an HIV-like disease that destroyed the bees' immune system.

In its first year, a CCD working group in the States, made up of scientists from six universities and led by the US Department of Agriculture, focused its efforts on trying to identify a virus or fungus. One team from Pennsylvania State University, Pennsylvania state department of agriculture and Columbia University made a breakthrough in September 2007 when it linked CCD with a virus that was identified in 96% of the hives affected by the disorder. But Israeli acute paralysis virus, which was first discovered in Israel in 2004, may prove to be a symptom rather than the cause of the problem. By recreating CCD in healthy hives, scientists hope to be able to determine what's triggering it.

With billions of dollars at stake, and the further expansion of the Californian almond crop in peril, the US Congress provisionally approved increased funding totalling around $100 million (£72 million) for research. But apiarists increasingly believe that the scientists are backing the wrong horse.

Dave Hackenberg, the beekeeper who is widely credited with discovering CCD in his hives in Florida, puts pesticides in the dock. He argues that bees have had viruses for years but a new type of nicotine-based pesticide is breaking down their immune system and causing CCD. Imidacloprid is his prime suspect. Not licensed in the US until 1994, it is now found almost everywhere from suburban lawns to apple orchards and sunflower fields. Bayer CropScience, the manufacturer, denies that its product is responsible for CCD and cites studies that support its conclusion, but other studies in France and Italy have found that the chemical disorientates bees, impairs their memory and communication and causes nervous system disorders. The French government was so concerned that it partially banned imidacloprid in 1999 pending further studies.

Many experienced beekeepers support Hackenberg's thesis, but scientists remain unconvinced. If pesticides are the culprit, they ask, why have bees disappeared from areas where none are used? Instead, they point the finger at beekeepers for overworking and undernourishing their bees. Hackenberg's 2,200 hives were logging 5,500 miles a year on the road before he lost two-thirds of them to CCD. In his defence, he says the trek hadn't troubled the bees before in all the 40 years that he's been doing it.

Bees have been disappearing long before pesticides or the stresses of modern life were invented. The first recorded unexplained loss was in the States in 1869, and thereafter large numbers mysteriously vanished in the US and Australia at intervals throughout the 19th century. Between 1905 and 1919, an epidemic wiped out 90% of the honeybee colonies in the UK. Throughout the 20th century, large-scale losses were reported throughout the States and in neighbouring Canada and Mexico. Then, as now, the main suspects were mismanagement by beekeepers, deficiencies in the bees' diet and chemicals in the environment, but the mystery was never solved.

Today's scientists are confident that, armed with many

new tools of detection, such as a complete mapping of the honeybee genome, they will be able to nail the culprit behind this latest outbreak. But more than a year after they began their investigations, they were still following leads.

Meanwhile, US beekeepers reported a second year of CCD. Hackenberg, who restocked after losing two-thirds of his bees in the winter of 2006-7, was dismayed to find that 80% of his colonies had vanished again when he opened his hives in Florida in November 2007. And results from an Apiary Inspectors of America survey revealed that 36% of colonies died over the 2007-8 winter, a 14% increase compared with the previous year. More than a third of beekeepers attributed at least some of the deaths to CCD-like symptoms.

If bees continue disappearing at this rate, it is estimated that by 2035 there could be no honeybees left in the US. In the UK in 2007, an estimated 25% decline in honeybees reported by the government's bee inspectors was not officially attributed to CCD but that didn't stop the then farming minister Lord Rooker from warning that all of the country's 270,000 or so colonies could be gone in 10 years.

There is a part of China where life already exists without bees. The uncontrolled use of pesticides is reported to have killed off honeybees in southern Sichuan in the 1980s. As a result, the pear trees have to be pollinated by hand: a slow, labour-intensive process that comes nowhere near to matching the bees' productivity, where one colony might take care of three million flowers a day. If such a process were tried in the US, it would cost an estimated $90 billion (£64 billion) a year.

In addition to fewer, and more expensive, fruits and vegetables in the shops, no honeybees equals no honey. Although it was migratory beekeepers who raised the alarm about bee disappearances, there are already reports of bee losses hitting honey production in Argentina, the world's largest exporter.

Scientists are now exploring a hi-tech solution to the vanishing of *Apis mellifera*. They want to engineer a new, virus-resistant superbee that would combine the hardiness of the aggressive Africanised honeybee with the docile nature of the western honeybee. While not beyond the realms of possibility in our world of cloned sheep and test-tube babies, a Frankenstein bee is not a panacea. If we put our faith in a hi-tech fix, we are ignoring the bees' environmental wake-up call.

We wanted to write a book that alerted people to the wonders of the honeybee and unravelled the mystery of its disappearance. In all the excitement, had some basic questions been overlooked? And were scientists, in their desperate search for a virus on which to pin the disorder, looking in the wrong place for the answers?

CHAPTER 1

IN THE BEGINNING
WAS THE BUZZ

The world's oldest bee fossil was found in a piece of Burmese amber believed to be 100 million years old. This first bee was much smaller than the western honeybee with which most of us are familiar, but it was covered with the same fluffy hairs to help it pick up pollen. However, unlike the modern honeybee, it is thought that this ancient relative was a solitary creature.

It is the honeybee's social behaviour, more than its ecological role, that has fascinated and amazed humans down the ages. How to explain the complex world of the colony, where workers gather and process food, care for the queen and her young, build nests and defend their community? No other creature has in turn been used as a metaphor for feudal hierarchy, absolute monarchy, republicanism, capitalist industry and commerce as well as socialist aspirations. The ancient civilisations of Egypt, Greece and Rome all paid homage to the honeybee's complex society, and primitive humans before them understood the insect's significance enough to bring honeybees to life in the earliest cave paintings.

We can trace the evolution of human society through bees, their honey and wax. The story begins somewhere in a cave in southern Africa, the birthplace of *Homo sapiens*. Honey for these cave dwellers was a real bounty, a sure-fire way of getting an instant calorie boost and the energy needed to hunt four-legged prey. Within the 4,000 rock art sites in Zimbabwe and South Africa, there are a dozen images of honey hunting dating back 15,000-20,000 years. The art shows bees in flight and a hunter smoking bees out of their nest, a practice that survives to this day.

The western honeybee migrated north and west from Africa, reaching Europe by the end of the Pleistocene era some 10,000 years ago, as the glaciers retreated and the last ice age drew to a close. Cave dwellers near present-day Santander in northern Spain had developed sufficient creative abilities to be able to make some wall paintings, probably as part of a shrine. The remarkable thing about the Bicorp painting, in a place known as the Araña or Spider Cave, is that it actually depicts a somewhat dangerous hunt for honey.

Its walls bear a painting of two human figures on a rope ladder ascending a cliff face. The figure at the top has one arm thrust into the cavity of a bees' nest, while the bees themselves hover around the entrance. His other hand holds what appears to be a basket, presumably to collect the spoils. There are many other such paintings of raids on bees' nests, dating from this time onwards. Their large numbers confirm that honey was highly valued by early humans. In fact, it was the original take-away. The first intoxicating beverage, meanwhile, was probably mead, fermented from honey.

Records indicate that the Chinese practised beekeeping more than 3,000 years ago, but it is the ancient Egyptians who deserve the title of the first beekeepers in history. The art of beekeeping was well established along the Nile by 2,400BC, and modern migratory beekeepers point to the Egyptians as the first people to transport their bees for honey production. This great civilisation had legions of scribes who wrote down every grain of produce before it was stored in state warehouses. The honey scribe certainly had his hands sticky. Records show that Rameses III offered up nearly 21,000 jars of honey to Hapi, the Nile god. A stone bas-relief from the Sun Temple, not far from Cairo, shows a figure kneeling beside a stack of clay hives and apparently blowing smoke into them to calm the bees. Five other figures are busy pouring and possibly straining honey from the hives into vessels of varying size, and finally sealing the jars. Containers of 3,000-year-old honey have been found alongside the mummified remains of a pharaoh, designed to sweeten the afterlife.

But honey was more than just a foodstuff to the ancient Egyptians – and to the civilisations that followed them. Honey was one of the ingredients used to try to prevent miscarriages, and was also applied in dressings for wounds, where it acted as an antiseptic. Beeswax was used in embalming, sealing coffins and mummification, and in candles to create artificial light.

The Pharaohs also used honey in their wedding celebrations. This custom was passed on to Greco-Roman culture and handed down to medieval Europe. Newlyweds drank honey wine for a month after the wedding ceremony to bring them luck and happiness. The ritual is one possible source of the word "honeymoon".

Temples kept bees in order to satisfy the demand for honey as offerings to the gods, while other uses included mummification, boat-building, paint-making and metal-casting. A papyrus recorded that when Re, the first of the Egyptian gods, wept, his tears "were turned into a bee" which "busied himself with the flowers of every plant, and so wax was made and also honey". The honeybee became part of the divinity of Egypt, and was an official symbol of Lower Egypt.

For ancient Greece, honey was both the food of the

gods and the stuff of love. In Greek mythology, Zeus – destined to be the god of gods – was nourished on honey as an infant, from which he derived his infinite wisdom. Wild colonies of bees gathered the best honey from their queens especially for him. The honey nymphs, or bee maidens, who fed the young Zeus were thought to be sacred because they saved him from being murdered by his father. Out of gratitude, Zeus gave the honeybee its sting, in unlimited usage, for its defence. Eventually a bee stung Zeus and he declared that "the bee abused its power" and decreed that the bee must die whenever the sting was used. Here was myth imitating reality.

About 2,500 years ago, the Greek poet Anacreon wrote about Eros, the god of sexual love and beauty, dipping his arrows in honey (the scene he described was later painted by the 16th-century German artist Cranach the Elder in a work called The Honey Thief). Anacreon's odes tell us:

Cupid as he lay among
Roses, by a Bee was stung.
Whereupon in anger flying
To his Mother, said thus crying;
Help! O help! your Boy's a dying.
And why, my pretty Lad, said she?
Then blubbering, replyed he,
A winged Snake has bitten me,

> *Which Country people call a Bee.*
> *At which she smil'd; then with her hairs*
> *And kisses drying up his tears:*
> *Alas! said she, my Wag! if this*
> *Such a pernicious torment is:*
> *Come, tel me then, how great's the smart*
> *Of those, thou woundest with thy Dart!*
>
> – Translated by Robert Herrick (1591-1674)

Another Greek, Aristotle, wrote extensively about bees in his natural history works. Though he did not realise that the ruler of the hive is a queen, not a king, in his History of Animals he described the birth of bees in the hive, the behaviour of the drones and of the workers, the way honey is collected, and gave details of early bee pests and diseases, as well as the bee's sting. And all this without a magnifying glass.

Classical Rome, in its turn, attributed the discovery of honey to Bacchus, the god of wine and intoxication, who also found time to show humans how to make beehives. The great poet Virgil wrote a practical beekeeping thesis in book IV of The Georgics, written from 36 to 29BC. He begins "Of air-born honey, gift of heaven, I now take up the tale", and advises farmers to "first find your bees a settled sure abode, where neither winds can enter" and a place where neither sheep nor goats "tread down the flowers".

Virgil vividly describes the workings of the beehive, from the role of the forager bees, "employed in getting food, and by fixed agreement, work on the fields", to the house bees' job of "laying their foundation of the honeycomb, then hanging the stickfast wax", and that of the guards, who "relieve incoming bees of their burden, or closing ranks, shoo the drones – the work-shy gang – away from the bee folds". He also says that bees "partake of an Essence Divine and drink Heaven's well-springs".

The Roman encyclopedist Pliny the Elder, meanwhile, called nectar the "sweat of the heavens" and the "saliva of the stars".

The ancients came to see the bee as a celestial creature that somehow filled the gap between human beings and the divine. After all, if nectar was a heavenly substance that fell from the sky, why wouldn't the creatures who collected it be sacred? "Some say that unto bees a share is given of the Divine Intelligence," wrote Virgil.

Like the Romans, Mexico's Aztec rulers practised beekeeping across the empire, and accepted honey as payment of taxes and tithes. This practice was also adopted by feudal kings in Anglo-Saxon Britain, where honey production became entangled in property laws. The Rectitudines Singularum Personarum, probably composed in the early 11th century, describes the various classes of tenants and functionaries of an Anglo-Saxon estate, their obligations and their privileges. Listed is a "beo ceorl" or beekeeper, who belonged to the lowest rank of free men, alongside the swineherd. A couple of centuries later, the Charter of the Forests declared that taking someone else's honey and beeswax was poaching. In medieval Russia, the Ordinance of Yaroslav, an early Russian legal code with 121 articles, had six laws protecting beekeeping. For damaging or burning a prince's hive, the fine was three grivna, which soared to 12 grivna "if anyone obliterates the sign on a beehive". At that time this sum was enough to buy six horses or 15 cows or 120 sheep. Damaging someone else's beard would also cost you 12 grivna!

At the heart of medieval society were the monasteries and abbeys. They were the largest owners of cultivated farm-land, had flocks of sheep and cattle, produced their own mead, honey and wax for candles – and sometimes extracted rents from their tenants in the form of honey. In religious terms, the bee was seen as an industrious, selfless worker for the greater good, who, in addition, was chaste – as it was then believed that bees reproduced asexually.

A popular theory for bee birth held that the insects were born from a carcass of a dead ox, nicely symbolising resur-rection and rebirth. This practice of "bugonia" has its most famous echo in the biblical story of Samson. When a swarm of bees make honey in a dead lion that the Hebrew has slain, it leads him to pose the following riddle to the Philistines:

"Out of the eater came forth meat, and out of the strong came forth sweetness." The incident is recalled on tins of Lyle's Golden Syrup, which picture a lion's head surrounded by swarming bees.

Honeycomb symbolised the cells where monks lived and worked. Bees were also regarded as a symbol of Christ, with the honey and sting representing his mercy and justice. Bees, it is said, hummed on Christmas Eve to honour Jesus at his birth.

Insect politics reached new heights in Elizabethan England. In Henry V, Shakespeare had the Archbishop of Canterbury speak about the division of labour in nature. He predicts harmony only when the social agents, King included, know their place in the hierarchy. Canterbury advises King Henry to consider the honeybees, which "in nature teach the act of order to a peopled kingdom":

They have a king, and officers of sorts,
Where some, like magistrates, correct at home;
Others, like merchants, venture trade abroad;
Others, like soldiers, armed in their stings,
Make boot upon the summer's velvet buds,
Which pillage they with merry march bring home
To the tent-royal of their emperor,
Who busied in his majesty surveys
The singing masons building roofs of gold,

> *The civil citizens kneading up the honey,*
> *The poor mechanic porters crowding in*
> *Their heavy burthens at his narrow gate,*
> *The sad-ey'd justice, with his surly hum,*
> *Delivering o'er to executors pale*
> *The lazy yawning drone.*

The apparent hierarchy and inflexibility of honeybee life inevitably became useful as a justification of feudalism and monarchy. In a beehive, after all, everyone knew their place and deferred without question to the ruler. Charles Butler's The Feminine Monarchie: or the History of Bees, published in 1609, celebrates the monarchy of bees as divinely ordained. Bees, wrote Butler, live "under the government of one Monarch" whom they "obey in all things". He approved of how the bees "abhor polyarchy, or anarchy", and all in all the beehive was a "pattern unto men".

The Pilgrim Fathers who founded the first English settlements in the New World sent for the honeybee soon after their arrival. In December 5 1621, the Virginia Company in London sent a letter to the governor and council in Virginia, saying that the Discovery had left England in November carrying "divers sorte of seed, and fruit trees ... and Beehives". Four arduous months later, the western honeybee landed in the New World. From Jamestown the honeybees multiplied and spread out, even though it was another 16 years before the next shipment made it to North America.

The role of the honeybee in the colonisation of America should not be underestimated. The settlers and their honeybees went hand in hand. Without the bees to pollinate the European seeds and saplings that the colonists brought with them, the land would not have been covered in the white clover that the imported livestock ate. As the poet Emily Dickinson later wrote: "To make a prairie it takes a clover and one honeybee."

In return, the settlers helped the bees over wilderness areas like plains and mountain ranges. It wasn't long before Native Americans understood that the sighting of the honeybee signalled death and destruction for them and their way of life. They came to refer to the bees as the "white man's fly".

In The Song of Hiawatha, written by Henry Wadsworth Longfellow in 1855, Hiawatha dreams of white men and his winged companions:

Wheresoe'er they move, before them
Swarms the stinging fly, the Ahmo,
Swarms the bee, the honey-maker;
Wheresoe'er they tread, beneath them
Springs a flower unknown among us,
Springs the White-man's Foot in blossom.

In an 1879 essay on bees, the American naturalist John Burroughs expanded on the theme. "The Indian regarded the honeybee as an ill omen. She was the white man's fly. In fact she was the epitome of the white man himself," he wrote. "She has the white man's craftiness, his industry, his architectural skill, his neatness and love of system, his foresight; and above all, his eager, miserly habits. The honeybee's great ambition is to lay up great stores."

Honey production rose sharply throughout the 17th century, and by 1730 Virginia was exporting beeswax to Portugal. It is estimated that the total beeswax exported that year was a staggering 156,000kg (344,000lb), which suggests that there were over 170,000 hives set up for the export market alone. Other records show that the honeybees themselves were valuable property. The average price per swarm was similar to the value of a sheep, and greater than that of a hog.

Eventually, the honeybees made it over the last geographic barrier – the Rocky Mountains. Some came by way of the Oregon trail; others reached California in 1853 by boat from Panama.

The honeybee became synonymous with America's struggle for independence. In 1779 the Continental Congress, America's first government, adopted as the logo on its currency a basket-like beehive with 13 rings – one for each of the colonies. The image was taken up by the Freemasons in the 1790s to emphasise American values. In 1838, the honey-bee became the official symbol of the Mormons' Church of Jesus Christ of the Latter-day Saints. The Mormon homeland in Utah was initially called Deseret, the Mormon word for honeybee, and to this day the state's emblem is a beehive.

Back in the Old World, the English Revolution led by Oliver Cromwell had put an end to feudal absolutism, and the self-organisation of bees had become synonymous with parliamentary rule, as shown in the title of Samuel Hartlib's The Reformed Commonwealth of Bees, which brought together a wealth of agricultural practices from around Europe. The first recorded use of the word "apiary" in English is from around this time. Hartlib's book explained how an estimated £300,000 worth of honey could be produced if the government systematically planned the setting-up of hives in every parish.

A new period of scientific understanding of agriculture and beekeeping had dawned since the Reformation. In Germany, Nickel Jakob wrote in 1568 that honeybees could raise a queen from eggs or very young larvae. In 1586, Luis Mendez of Spain was first to describe a queen as a female that laid eggs, and the mother of all bees. In 1609, Charles Butler showed that the larger drones were male; and in 1637 Richard Remnant showed that workers were female. But it was not until 1771 that Anton Janscha of Slovenia described the mating of a queen with a drone.

In 1717 the Frenchman Sébastien Vaillant demonstrated that nectar was produced by flowers. The scientific name *Apis mellifera* was given to the honeybee in 1758 by Carolus Linnaeus, the Swedish botanist who invented the system that is still used to name species. He got this one a bit wrong, however. His definition literally means "the honey-carrying bee". A more accurate name, *Apis mellifica*, or "the honey-making bee" was proposed in 1761, but by then it was too late.

In France, the struggle over the bee metaphor continued up to and beyond the revolution of 1789. Enlightenment thinkers like Jean-Baptiste Simon struggled to demonstrate that the hierarchy of the hive did not justify absolutism – quite the opposite. He called his book The Admirable Government, or the Republic of Bees. Others argued that there could be neither king nor queen in a beehive because there was no such thing anywhere else in nature.

So it was only natural for the revolutionary 18th-century French republic to adopt the honeycomb, in the shape of a hexagon, as its symbol, which also nicely coincided with the six-sided outline of France itself. When he was crowned emperor, however, the robe Napoleon wore for the ceremony was velvet embroidered with imperial bees in gold, in a brazen attempt to replicate the coronation of Emperor Charlemagne. Now the republicans' symbol was again in the service of dictatorship.

In the same year as the French Revolution, the Scottish utopian writer James Bonner wrote in The Bee-Master's Companion and Assistant of the hive as an enlightened society or "commonwealth" in which universally accepted order and law presides, "all are comfortable in their respective positions, and wealth and possessions are equitably distributed".

"The Bee called the Queen," he wrote, "so far as ever I could observe, has no sovereignty over the rest of the Bees." This was dangerous, seditious rabble-rousing that was quickly suppressed.

In 1819, the poet Shelley, in his "Song to the Men of England", used the bee metaphor to urge ordinary men to seize their rightful share of power:

> *Wherefore, Bees of England forge*
> *Many a weapon, chain and scourge,*
> *That these stingless drones may spoil*
> *The forced produce of your toil?*

Although there were stingless bees native to Australia, the western honeybee finally made it to the Antipodes in 1822 on the convict ship Isabella. Half a century after Captain Cook had mapped the east coast of Australia and, under instruction from George III, named it New South Wales, the bees arrived, somehow surviving the five-month voyage. In 1840, £4 was paid for a hive of bees in Jervis Bay, by a settler who hired an Aborigine to carry it 40 miles to his home. The western honeybee now had a whole new continent to explore. The Aborigines, like the Native Americans, associated the new arrival with the colonists. They called it "the white man's sugar-bag".

Not long afterwards, New Zealand had its first honey-makers. Housekeeper Mary Anna Bumby brought two straw hives of bees from Thirsk, Yorkshire, for her brother, who was a missionary. Two more hives came from Australia two years later. In 1844, the Reverend William Cotton left Britain for New Zealand, taking with him beehives ingeniously cooled with ice; unfortunately, the sailors thought the bad weather they encountered was caused by the bees and threw them overboard. Cotton, however, obtained bees once in New Zealand and wrote a manual for the country's beekeepers, as well as a treatise on bees in Maori.

By the early 19th century, honey was no longer the standard sweetening agent in Britain. The arrival of cheap sugar meant there was no need for every household in the land to have a hive in the garden. But as industrialisation expanded, the beehive, symbol of both hard work and the acceptance of the social order, still featured in Victorian society. Many buildings of the time had bees in their decoration, like Manchester's neo-gothic town hall, whose entrance hall features a mosaic depicting the bee. The insect is also carved into many of the pillars and walls.

George Cruikshank's 1840 print The British Bee Hive underlines the symbolism. His cross-section of the hive shows everyone in their allotted place, as part of the natural order of things. This was a common – and convenient – view in establishment circles. The Illustrated London News published a comment on the Great Exhibition of 1851 entitled "The Great Gathering of the Industrious Bees". The article spoke of "two hundred thousand little labourers ... diligently engaged in their various daily duties, while their reigning sovereign reposes quietly in her regal apartments."

For workers too, the earnest and self-motivated co-operation of bees in the hive served as a metaphor for their own communal interests. The first national trade union-sponsored newspaper, launched in 1862, was called The Bee-Hive. In November 1864, the paper became the official organ of the International Workingmen's Association, founded by Karl Marx. His own view of the usefulness of the bee metaphor is expressed in Das Kapital: "A spider conducts operations that resemble those of a weaver, and a bee puts to shame many an architect in the construction of her cells. But what distinguishes the worst architect from the best of bees is this, that the architect raises his structure in imagination before he erects it in reality." In other words, bees had their merits – but also their limitations by comparison with human beings, who could conceive of things in the abstract before putting their plans into practice.

With workers literally working themselves to death in the dark, satanic mills of northern England or on the vast steel railroads of east-coast America, the metaphor of the hive began to lose its attraction to radical reformers. But the workings of the hive continued to inspire in other fields of human endeavour. Architects such as Antonio Gaudí, who designed the Sagrada Família in Barcelona, Mies van der Rohe, founder of the Bauhaus movement, the American Frank Lloyd Wright and even Le Corbusier are said to have been influenced by the honeybee's practices. Juan Antonio Ramírez, in The Beehive Metaphor: From Gaudí to Le Corbusier, notes: "There is no doubt that a certain similarity can be seen between bees, which sculpt the honeycombs out of a soft substance (wax), and sculptors, who do as much with the same material (or equivalents) … Is it therefore so surprising that, in this context, the artist should be associated with the self-sacrificial bee?"

In Paris, La Ruche – literally the beehive – is an artists' residence in Montparnasse that has survived all the developers' attempts to destroy it. This three-storey, circular 19th-century structure got its name because it looked more like a large beehive than any dwelling for humans. The owner, Alfred Boucher, a firefighter and sculptor, wanted to help young artists by providing them with shared models and with an exhibition space open to all residents. Rent was nominal and artists came and went as they pleased. Few places have ever housed such talent as could be found at La Ruche. At one time or another, Marc Chagall, Fernand Léger, Amedeo Modigliani and Diego Rivera all called the place home or frequented it.

By the 20th century the western honeybee had been transported worldwide. Its exploitation was greatly facilitated by the invention of the moveable-frame hive, patented by the Reverend LL Langstroth in 1851. The new design – the first one ever was reputedly recycled from a champagne crate – contained a number of vertical frames in which the bees would build their wax honeycombs. They had previously been kept in upturned baskets or pots called skeps, which had to be destroyed in order to harvest the honey, but now the keeper could remove the frames at will, without killing the colony.

A final frontier for the western honeybee was China, with its venerable history of beekeeping. Honey heads the list of medicines described in the Book of Chinese Medicine, written 2,200 years ago. Even today, most honey produced in China goes towards the production of naturopathic remedies, the sweet liquid being mixed with medicinal herbs and turned into pill form. The ancient Chinese classified honey by the terrain from which it was gathered: as ground honey, bamboo honey, wood honey and even stone honey.

The country's native honeybee, the eastern honeybee or *Apis cerana* suddenly faced the invasion of the western honeybee at the beginning of the 20th century, when it was introduced from Russia. By the end of the last century, *Apis mellifera*, with its consistent and prolific rate of honey production, undeniably had the upper hand. For every native honeybee hive, there were six western honeybee colonies. They helped China to become the world's leading honey producer, with a 40% share of the global market.

The western honeybee's dominance is not totally unchallenged, however: in the US it is threatened by the Africanised honeybees that are a hybrid of African honeybees that were introduced to Brazil in the 1950s. These "killer" bees, as they have been named because of their aggressiveness, have spread north and south, displacing the western honeybees they meet on their way.

Now, at the start of the 21st century, the honeybee is making one final journey: into the laboratory. In 2006 scientists unravelled *Apis mellifera*'s genetic code; in time, they claim, their work will benefit agriculture, biological research and human health. So far, the geneticists have gone some way towards explaining the bees' perfect governance and efficiency.

Over tens of millions of years, the bees' genes have created dramatic distinctions in shape, physiology and behaviour within the hive. Honeybees secrete chemicals known as pheromones to reinforce individuals' place within the social order by broadcasting their gender, age and caste.

Since its split from the larger and rounder bumblebee 60 million years ago, the honeybee has evolved the highest rate of recombination – the process by which genetic material is mixed during sexual reproduction – of any known animal. This is thought to boost genetic diversity among honeybees, where the queen is the only reproductive female in the colony.

The scientists also discovered that, like humans, bees have genes that encode the 24-hour "circadian rhythm", giving them an innate sense of day and night. This probably makes them more efficient at gathering food, just like the relatively high number of genes that encode their sense of smell.

But it's not all good news for honeybees. Geneticists have also discovered that they have fewer genes providing resistance to disease than other insects. The size of the major gene families responsible for detoxification also appears to be smaller in the honeybee, which could make it unusually sensitive to pesticides and poisons.

Despite millions of years of evolution, might the honeybee be ill-equipped to cope with the man-made forces now being thrown at it? Do bees' inherent vulnerabilities go some way towards explaining their mysterious disappearance?

CHAPTER 2

THE MASTER POLLINATOR

On a warm, sunny April day a honeybee is attracted to a patch of bluebells. She – for almost all bees are female – is on a mission, and knows that she is close to her goal. She delicately clambers over the petals, as comfortable upside down as she is right way up, her six legs working in unison. She pokes her head inside a trumpet-shaped bloom, searching for the sweet, sugary nectar.

Grasping the petal's side, her legs force her head deep inside the flower, until her long tongue can get to the liquid. Once the tiny drop is all gone, she will move on to the next flower, and the next until she is full.

While the honeybee is busy lapping up nectar, she is also unknowingly carrying out one of nature's most important tasks, fertilising the flowers as her hairy body transfers pollen

– the sperm cells – from the male parts known as anthers to the female part called the stigma. At the base of the stigma is the plant's ovules, where the egg cells are fertilised by the sperm. This is why nectar is placed deep in the flower: to provide the best chance for the pollen to be picked up by the honeybees' electrically charged hairs.

When the bee stops on top of a petal, she may look as if she is brushing off the pollen grains, but in fact she is gathering them together with her middle legs, before stuffing them in basket-like structures on her hind legs and eventually carrying them back to the colony.

The honeybee may play a vital role in the flower's reproductive cycle, but she is also a beneficiary. The carbohydrate-rich nectar and the protein-packed pollen are the only natural food sources she can digest. She is therefore totally dependent on the flowers she visits.

Which came first: the nectar or the pollinator? We may never know, but what we can be certain of is that the honeybee and the pollen producers have evolved to work in perfect harmony over millions of years, each for their own gain. She has been designed specifically for this role.

To find her way around, the honeybee is equipped with not one but two sets of eyes. On the front of her head are two large compound eyes, each made up of 6,900 hexagonal lenses that can all act independently to decipher light conditions, colours and the position of the sun. These eyes are particularly good at picking up movement and have hairs at the junction of the lenses that are used in determining wind speed and direction. They are also sensitive to ultra-violet light – invisible to humans – expanding the already vibrant colours of the plant world into an explosion of differing hues, and directing the honeybee towards the area of the plant where the nectar and pollen are stored. To us a primrose may look a uniform yellow, but the bee sees a difference between the outer petal and the pollen-rich centre, with lines directing her to the heart of the flower in much the same way that the lights of an airfield direct planes to the landing strip.

The bee's other eyes, known as ocelli, are set out in a triangle at the top of her head. These three organs function as light detectors, to keep the insect the right way up.

The bee is not reliant on sight, however. Two antennae on top of her head act as smell detectors, and are particularly attuned to useful odours. She is up to 100 times more sensitive than humans to odours such as flowers, nectar, wax and propolis, the gluey resin that bees collect from trees and plants to fill gaps around their homes. Because each antenna picks up odours independently, a bee can locate a smell by turning her head until the intensity is equal for each of them, in much the same way that humans locate sounds with their ears. The bee, in other words, smells in stereo.

In order to reach deep into flowers, she has a proboscis,

or folding tongue, that can reach about half her body length when extended, while her rear pair of wings can fold back on to her body to enable her to enter narrow spaces.

Instead of teeth, her mouth has two powerful mandibles that work like pincers to hold, pull and bite, as well as manipulate wax and collect propolis.

Her feet, meanwhile, are equipped with both hooks for hanging on to the edges of petals and wax comb, and pads that allow her to walk upside-down on flat surfaces. Her front legs have specially adapted hooks that can also be used to clean her sensitive antennae, while the middle legs collect the wax that she produces underneath her abdomen and pass it to the front legs for manipulation by the mandibles when she is making comb. They also brush the pollen into the hind legs' baskets, or corbicula, which can each hold up to 8mg ($\frac{1}{3,500}$oz) of pollen while the bee flies between flowers.

That journey, by the way, will take place at a near-constant 24kph (15mph), as the wings beat 230 times per minute. Unlike other flying creatures, a bee that is carrying a load does not increase the rate at which her wings beat to gain extra power, but instead increases the distance that the wing moves through the air.

To defend herself, she injects a venom known as apitoxin through a barbed sting at the end of her body. Theoretically this sting can be used more than once against other bees, but it lodges in mammal or bird skin, leading to the bee's death as it is ripped out of her.

The honeybee engages all these parts of her 12mm-long (½in) body to do what she does best: collect nectar and pollen from a vast range of flowering plants to sustain her colony. As we have seen, she also performs a vital service for many of the planet's plants.

Different plants have devised different solutions to the problem of reproduction. Some employ the wind to blow the pollen off a male flower in the hope that it will land on a female flower of the same species. These anemophilous (literally wind-loving) plants produce great quantities of lightweight pollen grains, sometimes with air-sacs to keep them buoyant, since this scatter-gun approach means that most of the pollen will be wasted.

A much more efficient method is used by the 80% of flowering plants that engage the services of pollinating agents such as bees, butterflies and hummingbirds. Whereas anemophilous plants such as corn, wheat and rice have plain, inconspicuous flowers, entomophilous (insect-loving) plants have to attract their go-betweens with brilliantly coloured and strongly scented flowers that contain a rich treasure.

This system of reproduction may be more direct than pollen grains swirling around in the wind, but it is by no means foolproof. Having a variety of animals randomly visiting your flower is no guarantee that the pollen will be transferred to the female of the same species. What happens if an animal feeding on a bluebell's sweet juices decides to take its next drink from a tulip? The pollen on its body will be wasted. And this flitting from one type of flower to another is characteristic of opportunistic pollinators such as butterflies and bumblebees.

Some plants have forged a relationship with only one type of insect. These "monolectic" animals, such as solitary bees, do a great job for their symbiotic partners, whose nectar and pollen are the only food and drink they consume. It is almost 100% certain that if they visit a plant, its pollen grains will be transferred to the female parts of the same species.

The majority of plants, however, rely on "polylectic" insects like butterflies and honeybees. The honeybee collects pollen from a wider range of plants than any other known pollinator. In their book The Forgotten Pollinators, Stephen Buchmann and Gary Nabhan report on a study in the Santa Catalina Mountains in Arizona that found in five honeybees hives' in one year pollen grains from 55 flowering plant species, a quarter of all the kinds of plants blooming within the bees' foraging range.

Isn't this bad news for plants that rely on honeybees to spread their pollen? No, because a honeybee makes thousands of visits to one flower species at a time until the food source has dried up.

The work demanded of the foraging honeybee is truly astounding. She will visit 1,500 flowers to collect just one load of pollen, which will weigh about 15mg ($\frac{1}{1,900}$oz), about half as much as the nectar that she also brings back to the hive. To put these figures into some kind of perspective, it takes two million trips by a colony to collect the 30kg (66lb) needed to raise its young, and four million trips to collect enough nectar to turn into honey for winter stores. This equates to around 45,000 trips per day per colony. Since a foraging flight may take a bee on a 10km (six-mile) round trip, collectively a colony can fly up to 450,000km (280,000 miles) a day. Each bee will fly around 800km (500 miles) in her lifetime, at times carrying loads equivalent to half her body weight; no wonder she will die of exhaustion about three weeks after her first flight.

It is not just the bee's anatomy that has evolved to permit these amazing feats, however. The honeybee has also developed a sophisticated communication system, a celebrated dance that conveys distances and direction.

A returning forager laden with booty wants to tell her sisters where to find the rich pickings so that as many bees as possible will collect from the most profitable source of food. But a bee has no ears, and can not communicate by sound, so she uses first smell, then the vibratory "waggle dance" originally brought to the world's attention by the Nobel prize-winner Karl Ritter von Frisch.

In the complete darkness of the hive, the forager lets her sisters taste and sniff a piece of her treasure, before beginning to beat out a tune by shaking her abdomen from side to side as she walks up or down the face of the wax comb. The amount of waggle, or shake, tells them the flying time to the target: if she waggles her body for just a fraction of a second, it is nearby, while if the shaking lasts for a couple of seconds, the source is five minutes away. The dance will even reflect the wind speed and its effect on the length of the journey.

To show the direction of the food source, the dancer gives a bearing relative to the sun. When the food is directly towards the sun, she will waggle her way directly up the face of the comb. If it is in the direction opposite the sun, she will dance with her head pointing directly down the face of the comb. Still more remarkably, her dance will compensate for the movement of the sun, which is one degree every four minutes. So if at the time she arrived at the hive the food source was directly in line with the sun but it took her four minutes to be able to tell her news, she will waggle one degree off the vertical. She will repeat her message a number of times until she is happy that her audience has understood her and she has recruited a few more foragers to her source of food.

Each forager that returns from a successful search performs on the dance floor and attracts a following of her own; the greater the quality and abundance of a food source, the more foragers will be doing the same dance. If a forager feels that some of her sisters are employed on other, less important, tasks she will venture further inside the hive and dance a second routine, called a "shaking dance", where she holds on to one of her sisters by the forelegs and shakes her abdomen. This stimulating dance will hopefully encourage more bees out on a foraging flight. If, however, there is a shortage of bees to help foragers unload their baskets when they return to the hive, the foragers will engage in a "tremble dance" to draft in more recruits.

It is the sheer number of honeybees that can be mustered that puts this bee head and shoulders above others in the pollination stakes. With a colony numbering up to 50,000 at the height of summer, it is not unusual to have 20,000 honeybees all busy foraging on one type of plant. Compare this with a bumblebee colony of a few hundred, and you can see why the honeybee is so valuable to the plant world.

Within each colony, all of the bees will be the offspring of a single female, the one we describe as the queen. But that royal title is misleading: she is more of a slave to the community than its ruler. Her purpose in life is to lay eggs and populate the colony, but it is the workers that determine when, where, and how many eggs she lays. She is reliant on her daughters to feed and clean her and tend to her needs, and since her egg-laying rate is determined by the amount of food she digests, her workers have the balance of power. When the colony needs to expand, the queen is fed copious amounts of food, which triggers her laying. She can lay 2,000 eggs a day, round the clock. They look like grains of rice.

After three days the white larvae hatch from the eggs and begin six days of intensive feeding: three days solely on royal jelly – a white, protein-rich substance produced by the bee's hypopharyngeal gland – followed by three days of a pollen-honey mixture. The larvae rapidly grow and go through a series of instars (moulting of the outer skin), as they get bigger. When they stop eating the workers cap the cell with a wax covering, and the larvae spin their cocoon and spend the next 12 days undergoing the process of metamorphosis – changing from pupae into bees. Worker bees will emerge 21 days after the eggs have been laid, and drones three days later. Together, a colony's eggs, larvae and pupae are called the brood, and the area within the hive where they are reared is known as the brood nest.

When the queen begins to lose her fertility, usually after two to three years, it is time for her to make way for a new egg-layer. The workers will create a special queen-rearing chamber around a newly laid egg and when it hatches into a larva feed it plentiful supplies of royal jelly. This substance activates a change in the normal development of the larva and turns it into a queen equipped with ovaries, queen pheromones and a barbless stinger. When a new virgin queen bee emerges after 16 days in the cell and is mated, the worker bees will have no qualms killing their clapped-out old mum. This is called supersedure and is one of two ways that a colony "re-queens". The other method involves swarming.

When a colony decides to swarm to reproduce, usually in early summer, it starts by raising a number of queens, one of which will take over the hive while the existing queen flies off to find a new home. The workers do this by building a number of larger queen cells and either forcing the queen to lay an egg in each one or transporting in newly laid fertile eggs. Again, feeding the larvae lots of royal jelly turns them into queens.

Before the virgins emerge from their cells, the old queen will leave the nest with all of her young foragers in tow, leaving behind the older foragers and the house bees – those members of the colony that are not yet old enough to forage and instead take care of tasks such as cleaning the hive, looking after the brood and receiving nectar and pollen. The travellers gorge on a huge breakfast, equivalent to three days' worth of food, to see them through their quest for a new home.

A swarm is a spectacular thing to observe. The air around the nest is filled with the swirling of thousands of bees pouring from the entrance. They are searching for their queen and looking for a place to settle before they embark on their journey to their new home. This excitement subsides when the air becomes saturated with the colony's pheromones, beckoning the bees to congregate. Within less than an hour the spread-out cloud of bees has formed a tight mass, the size of a large melon, around a nearby tree branch with the queen securely held in the centre. Scout bees will have already been searching for a suitable new cavity for the swarm to occupy. Not long after the swarm has settled, they will encourage the throng to take flight and lead it to the new nesting site. There the colony will begin a new life by quickly building wax comb to hold eggs, pollen and nectar. In just 21 days, when the newly laid eggs hatch, the new colony will be fully functioning.

Back at the original home, the young house bees and the older foragers wait for the new virgin queens to emerge from their cells. Since the colony can accommodate only one queen, the virgins must fight to the death. The victor then needs to find a mate. She does this by taking flight outside the hive, high in the air, wafting her pheromones to attract a number of eager drones – the male bees. She will mate with a number of them in this nuptial flight to ensure the diversity of the gene pool and can store the sperm in her body for many years. Mating is the sole reason for the drone's existence, and his cause of death. When he does the deed, his genitals are ripped from his body.

Honeybees may be merciless, but they are fastidiously clean and tidy creatures. They groom each other and keep their home spotless. Dead bees are quickly dragged out of the hive and deposited outside, and bees are careful not to defecate indoors, even during winter, when they will wait for a sunny day to dart outside for a toilet stop. Such hygienic behaviour stops the spread of disease. Sick bees will always try to leave a hive so as not to spread illness throughout the colony.

Unfortunately, this instinctive desire to keep the home clean can sometimes backfire. If, for example, a colony is infected by nosema, a single-cell parasite found in the bees' gut that causes dysentery, the bees will defecate inside, and the only way that healthy specimens can clean up the mess is by using their mouths. In the process they ingest the parasite, and the disease spreads. Fortunately, a strong colony with a healthy queen has a chance of riding itself of the infection by laying new healthy brood to replace the sick.

In addition to the size of the workforce, the reason that honeybees collect so much more nectar and pollen than other pollinators is that they live through the winter. Solitary bees and even colonies of bumblebees die at the end of each summer. The only survivor is the bumblebees' newly mated queen, which hibernates until spring, at which point she finds a nest to lay her eggs and create a new colony.

But honeybees will only survive the winter if they have stored enough honey to prevent the colony from starving in the cold months ahead. So they waste no time turning the nectar they collect into winter stores. Even before a forager gets home, she starts the process by adding enzymes to the nectar to begin its transformation from the complex sugar sucrose to the simple sugars fructose and glucose. Once back at the colony, she regurgitates the nectar onto the proboscis of a house bee, which adds more enzymes and evaporates some of the water, like a chef reducing a sauce. The house bee then deposits the liquid into a cell, where other bees join in the honey-making process, using their wings to evaporate the water from the nectar. When the water content has been reduced to around 18%, the nectar has been successfully transformed into honey. When the cell is full, it will be capped with a fresh layer of wax. The honey is now in an airtight container, ready to be eaten when needed.

Pollen is also taken off the foragers when they return to the hive. The pellets are moistened with regurgitated honey and saliva, and packed into cells in the nursery near to the larvae by house bees using their mandibles and forelegs. The enzymes and honey added to the pollen turns it into what is sometimes called "bee bread" and stops it from fermenting. The protein-rich food can be stored for many months and is fed to larvae during their development and eaten by young worker bees during the first 10 days of their life, when their glands and internal structures are growing.

One of the reasons honeybees need so much honey to see them through the winter is that they expend a lot of energy just to keep their home heated, a process that involves them clustering together and shivering. The hive must be kept at 35C (95F) in order to incubate the larvae. The bees will usually manage to maintain this ideal temperature in the centre of the nursery, but if there are too few adults the fringes of the brood nest area can get chilled and the brood die. This may also occur because of a draughty hive, or a hole created by an animal in search of food or shelter.

Spring can be a precarious time for a colony, since the changeable weather can play havoc with the balance of the hive. A warm late winter and early spring will give the signal for the colony to expand, but if the temperature then suddenly plummets there will not be enough house bees to keep the newly laid eggs warm. If the temperature stays low for long, the colony may eat its way through its stores and, unable to go outside to forage, starve to death.

All this action takes place in and around the complex of six-sided cells that contains the bees' birthing chambers, pantry, warehouse, bedroom, offices and dance floor. The honeycomb is one of nature's most accomplished feats of engineering. As Darwin said: "He must be a dull man who can examine the exquisite structure of a comb, so beautifully adapted to its end, without enthusiastic admiration."

"Beyond this stage of perfection in architecture, natural selection could not lead," he added, "for the comb of the hive-bee, as far as we can see, is absolutely perfect in economising labour and wax."

In the wild, when a colony of bees finds a new cavity in which to live, the workers immediately start building six or so vertical sheets of wax, from the sides of which sprout thousands of horizontal hexagonal cells of uniform size. Each sheet is spaced exactly the correct distance from each other so that the bees can pass by each other back to back without getting squashed. This is the structure that apiarists attempt to recreate in their moveable-frame hives, so that sheets of combs can be manipulated without damaging the colony.

Once the honeycomb is built, inside the darkness of the hive, the bees work together in harmony, making sure the nest is always kept at the optimum conditions. The household tasks are closely related to the age of the bee, reflecting the physical changes as she matures. For example, when the bee emerges from its cocoon she is not completely formed. The soft exoskeleton takes a few days to harden, and the sting, wax, mandibular and hypopharyngeal glands are not yet developed. In her first few days, therefore, she spends her time feeding herself and cleaning cells. She then feeds the older larvae with pollen and honey until her hypopharyngeal glands are developed enough to feed the younger larvae a little royal jelly. Once her wax glands are secreting, she can begin to cap the cells of the pupating larvae and carry out repairs to the comb. As she nears her 14th day her sting is formed and she may be required to go on guard duty at the door of the hive. Towards the 21st day, her hypopharyngeal glands have enough enzymes for nectar processing and her more streamlined body is ready to take her foraging.

Despite this age-related specialisation, the colony can at times call on its inhabitants to take on any task it deems fit. So if ventilation, for example, becomes an issue, any of the worker bees can be called upon to circulate the air. The same is true for comb-building or warding off unwelcome visitors such as wasps during late summer, or mice during the winter.

Pheromones also play a part in assigning work. Tests have shown that interfering with these hormones can alter bees' activities, with damaging results. For example, young bees may embark on foraging trips before they are fully prepared and desert their nanny duties. Unable to feed themselves, the larvae will starve.

To ensure all the foraging bees get home after a hard day collecting nectar and pollen, those that are still in the hive emit a pheromone that says "Here we are!" to guide their sisters back. This is the same pheromone that is used when the colony swarms and congregates before flying off to their new home.

A honeybee knows which nest to go home to by mapping out its locale. The accuracy of its measurement is quite uncanny, and when a number of hives are located near to each other with the entrances separated by a distance of only a few tens of centimetres the honeybee will return to its own colony, rather than the one next door.

For the commercial apiarist whose business is crop pollination, this homing instinct is invaluable. When his bees have finished pollinating an orchard of apples he can wait till dusk – when the foragers have returned to the hive – and safely move the hive to a field of oilseed rape miles away. When the bees leave the hive the next morning, they realise that their surroundings have changed and redraw their map with the new position of the hive pinpointed. The only time this fails is if the hive is moved less than 4km (2.5 miles). The bee does not remap its location when the hive is moved a short distance, so the forager will return to the original location and search for hours for the missing colony. It will likely die as night falls and the temperature drops.

This ability to transport colonies of western honeybees without seemingly disrupting their lives is a major reason that they have become the earth's number one managed pollinator. But has it also encouraged a dangerous tendency to rely on them to satisfy all our pollination requirements, to the detriment of other insects, and to concentrate on just a few, increasingly vulnerable bloodlines?

CHAPTER 3

THE SHRINKING GENE POOL

The small, windswept island of Læsø lies in the Kattegat strait between the Danish peninsula and Sweden. Moors, heather and meadows blanket its interior and sand dunes hug its rugged coastline. A couple of thousand people live on the farms, villages and small ports scattered along the shore. In 1993, Læsø became the setting for an ambitious experiment to try to save the biodiversity of the western honeybee.

Millions of years ago *Apis mellifera* is thought to have spread northwards from Africa to occupy most of Europe; its territory stretched from the Cape of Good Hope to the Urals. Over this vast landscape habitats and climates change dramatically – from the harsh winters, late springs and hot summers of eastern Europe, through to temperate, maritime conditions in the north-west, and the dry desert terrain to the south. The bee adapted to these different environments through natural selection, producing more than 20 subspecies or races. These can be grouped into four evolutionary branches: an African branch, an Oriental branch, a north Mediterranean branch and a west European branch, each displaying different physical and behavioural characteristics. Studies have identified 36 physical differences between races of western honeybee, ranging from overall size, to colour, hair length and the pattern of veins on the wings.

In Europe, the bees survived the last ice age on two refuges: one on the Iberian peninsula and the other on the Balkan peninsula. After the glaciers retreated, the west European branch occupied north and west Europe, and the north Mediterranean race occupied Italy and the Balkans.

Since the 20th century, however, this natural geographical distribution has been seriously disturbed by human activity. Two races in particular from the north Mediterranean branch of the family have been introduced across the continent. The gentle and productive *Apis mellifera ligustica*, from Italy, and *Apis mellifera carnica*, from the former Yugoslavia, now dominate European beekeeping. As a result, some of the 20 subspecies are considered endangered. An example is the black bee (*Apis mellifera mellifera*) from the west European branch of the family, which once spread from Spain up through France and Germany across Russia and Britain and Scandinavia.

In Denmark, beekeepers favoured the Italian bee because it was better at pollinating red clover and produced copious amounts of honey. But a few survivors of the black bee clan held out 22km (14 miles) off the mainland.

In 1986, the first attempt to identify the Læsø bee population was carried out by the Danish Institute of Plant and Soil Science, which tested bees from 116 colonies in 27 apiaries. The colour of the bees and the pattern on their wings suggested that 97% of the examined bees were the original settlers – black bees. Only two colonies were categorised as Italian bees or hybrids. In 1990, a similar survey of 151 colonies from 20 apiaries confirmed that most bees were from the black bee race (although Læsø bees are actually brown in colour). By using a technique to trace maternal lineages among the black bees of Europe, researchers discovered that the Læsø variety was unique to Denmark. Genetic studies also showed that it had a distinct DNA make-up.

However, a creeping hybridisation of bees over the four years between the studies was also observed. In response, the Danish Beekeepers Association and its Læsø affiliate launched a scheme to try to stop the local bee from mating with outsiders. They set up a controlled breeding area on the island, and in 1989 bee imports to the island were banned. But not all the island's beekeepers agreed with the conservation measures.

Ditlev Bluhme was one of 10 professional beekeepers who flouted the ban. He claimed his Italian bees had greater resistance to illness and were much more productive than black bees, so switching to Læsø bees would seriously affect his livelihood. With agriculture and tourism the only sources of income on the island, he argued that beekeeping was one of the few lucrative trades open to him.

Since the scheme's inception, objectors have been engaged in a bitter dispute with the conservationists and tried to sabotage their work. The battle turned nasty after the Danish government signed up to the 1992 Rio convention on biological diversity and toughened the country's beekeeping rules to preserve threatened races. New regulations authorised the agriculture minister to outlaw the keeping of bees, except black bees, on Læsø. Swarms of other bees could be removed or destroyed, or their queen replaced with a Læsø queen. Following the introduction of the controversial new laws, there were reports of black bee apiaries on the island being ransacked by objectors.

Although the state agreed to pay full compensation to the owners of non-native bees, Bluhme and his supporters refused to give up their Italians. When the government prosecuted one of the law-breakers, Bluhme took his case to the European Court of Justice.

Bluhme disputed that there was a pure strain of black bee unique to Læsø to be saved. He argued that the island's bees were still found in many parts of Europe and were not at risk of extinction. And he accused the Danish government of restricting free trade by refusing to allow him to import Italian bees.

In its defence the Danish government cited the Rio convention, which affirms that the "conservation of biological diversity is a common concern of humankind" and should be applied even to "species in which the evolutionary process has been influenced by humans to meet their needs". The convention endorses preventative measures to preserve biological diversity, such as establishing protected areas where invasive alien species can be controlled or eradicated.

In December 1998, the court backed the Danish government, ruling that it was legal to protect threatened subspecies under article 36 of the European Community treaty, which deals with the protection of the health and the life of animals. This was the case even if it meant banning imports of other species.

The islanders of Læsø have a reputation for being fiercely independent of mainland Denmark and of treating the authorities with suspicion. Bluhme's dispute was the latest in a long line of grievances between locals and officials from Copenhagen stretching back centuries. And he was not going to be deterred by a decision by the European Court in

Luxembourg. Bluhme refused to give in, and it was another three years before he lost a final appeal in a national court.

Illegal colonies of Italian or hybrid bees persisted on Læsø as apiarists continued to break the law and threaten the conservation efforts of 26 beekeepers, who between them kept 150 colonies of native bees. In 2005, a working group was set up by the Danish Plant Directorate to mediate a settlement between the warring factions. It decided in the interests of human liberty to limit the ban on imported bees to a small area on the east of the island.

Conservationists feared this compromise would fail to keep the population of bees apart. A survey in 2003 had found that only 35% of the island's black bees were of pure race. But a 2007 survey identified a total of 123 colonies of black bees, less than 1% of which had genes of another race. "The protection areas have proven of value for the conservation of the *Apis mellifera mellifera* bees," says project leader Per Kryger, a senior scientist at Aarhus University's faculty of agricultural science. "Not least due to the local acceptance and respect of the new regulations."

Additional islands are now being proposed as conservation areas for the black bees. A group of Swedish beekeepers who visited Læsø to collect queen eggs and drone sperm to create an overseas breeding programme described the native bees as good-natured, hardy and healthy. "Losses during the winter are almost unknown. Nosema illness does not exist.

They build fantastic combs," Ulf Groehn wrote in his account of the visit. The only downside he reported was that they "swarm a little".

Professor Robin Moritz, who led a European Union-funded research network on Beekeeping and Apis Biodiversity in Europe (Babe) from 2001 to 2003, argues that the genetic diversity that the Danish Beekeepers Association is trying to conserve on Læsø is of great importance for the continued use of the honeybee in agriculture. "If we create a pan-European bee it will be difficult to get back to the original subspecies that had adapted to particular habitats and climates," he says. "Babe was about keeping subspecies, as their different traits may come in handy when we need them."

A homogenous model is predominantly used across Europe, however. More than 55 years ago German beekeepers selected the gentle-natured and productive *carnica* honeybee. And Italian immigrants filled the void when most of Britain's native honeybees were wiped out by a virus a century ago. DNA analysis of the French honeybee population shows that recent arrivals from Italy and the Balkans have already been incorporated into the local gene pool.

Moritz points out that these newcomers are not so well suited to northern Europe's long winters. Black bees are typically better able to cope with cold climates. They wait until late spring, when temperatures have risen, before they expand their numbers, and their emphasis is on nectar collection from late-flowering plants to see them through the winter. A bee adapted to colder climes, he says, may fare better at fighting off fungal infections that germinate when bees are confined to their hive for months at a time.

Bees in cool climates, at increasing geographical latitude or elevation above sea level, are also typically bigger, stouter and darker than their cousins from warmer climes. The *Apis mellifera iberica* native to Portugal and Spain, for example, is generally smaller and shorter-haired than northern European honeybees.

But the selection process practised by commercial beekeepers has traditionally tried to produce bees that are docile, for the amateur beekeeper, and productive, for the honey producer, and build up rapidly in the spring to exploit early honey flow, for the commercial pollinator. This has unavoidably led to a reduction of genetic diversity.

"Commercial selection is directly opposite to what natural selection would achieve," says Moritz. By breeding a gentle, efficient honey-maker, he warns, we have made bees much more susceptible to disease than they would be if natural selection had played a role.

He cites an experiment on Gotland, an island in the Baltic Sea, where bees not treated for the blood-sucking varroa mite – which has caused huge losses of colonies in

Europe and the US – adapted naturally over six years to control the parasite. Although most of the 150 colonies died, winter mortality rates decreased from 76% in the third year to 19% in the sixth as mite infestation fell. Similar results were observed on a tropical island off Brazil where mite-infested European bee colonies were left untreated since 1984: after an increase in infestation levels, mite numbers fell, leading to a balanced relationship between host and parasite.

The bees that survived on the Swedish island live in very small colonies, have a short brood cycle and produce few drone brood. The reproductive success of varroa mites is higher when more immature drones are available for them to lay their eggs in. Ingemar Fries, entomology professor at the Swedish University of Agricultural Sciences, is careful not to claim that the Gotland bees are varroa-resistant. Instead he says: "Our experiment demonstrates a possible host-parasite co-adaptation, ensuring survival of both the host and the parasite. We don't know if Gotland bees are equally tolerant to other mite populations." Initial studies in Germany suggest they might be, but ongoing work with Swiss and Swedish colleagues will find out.

In contrast to the Gotland bees, the European apiarist's bee of choice has long, populous brood cycles that favour varroa's reproduction. Moritz believes the predominance of the Italian bee is a major reason why the mite has proved so lethal across the continent. "Without a doubt," he says, "commercial selection has allowed the mite to flourish."

In an academic paper, Fries and his colleagues say that the results on Gotland so far allow them to conclude that the problem facing the apiculture industry with mite infestations is probably linked to beekeepers controlling the mites with chemicals and thereby removing the bees' own selection process.

Wild bees that lived in Europe's forests used to provide an important check against the loss of genetic diversity. The honeybee is the only agricultural animal where the managed population regularly shares its gene pool with the wild population. Young queens try to mate with as many suitors as possible, some of which will have flown up to 10km (six miles) to join the party, lured by the females' pheromones. The more drones that participate, the less risk there is of a queen having sex with a brother. She can mate with up to 20 drones in mid-air, storing their sperm for two to three years, so there can be 20 paternal bloodlines, or patrilines, in one hive, each with a slightly different genetic make-up.

Yet despite the queens' polyandry and the honeybees' enormous rate of recombination – the process by which genetic material is mixed during sexual reproduction – they are unable to produce a varied gene pool if there is not a diversity of genes in the population.

"The most dangerous but often disregarded risk for biodiversity stems from excessive gene flow from vast numbers of managed colonies into a decreasing wild population," says Moritz. "A massive importation of honeybees can lead to the newcomers' gene flow destroying the genetic architecture of the local race."

But wild western honeybees faced an even more serious onslaught with the arrival of the varroa mite 30 years ago. Unable to defend themselves against the new parasite, most perished, leading to fears that there would be a drastic loss of genetic diversity, creating what scientists call a population bottleneck.

It was against this background that Babe was set up to bring together honeybee geneticists from across Europe to study ways of conserving any indigenous bees that had survived. It was hoped that by increasing genetic variation higher resistance to disease could be achieved. Babe urged commercial beekeepers to source their bees locally, from native races, rather than relying on pre-selected stock obtained somewhere else on the globe.

Secluded mountain areas or islands were identified as the ideal locations for conservation breeding programmes. Mountains and long stretches of water act as a barrier to gatecrashers. But weather conditions also need to be good for successful mating. Virgin queens have just a few weeks after they are born to copulate. If it's too cold or wet to leave the hive, they will miss their opportunity and the workers will replace them with newer models.

EU regulations are in place to support the development of native species. The directive on organic beekeeping is based on using indigenous honeybees rather than imported stock. The directive says "a wide biological diversity should be encouraged and the choice of breeds should take account of their capacity to adopt to local conditions ... Preference is to be given to indigenous breeds and strains ..." But there are fears that organic apiculture is impossible without restrictions on bee movements to prevent uncontrolled gene flow.

In the US there are no native honeybees to conserve. Before the Pilgrim Fathers arrived in the 17th century there were thousands of native pollinators, including at least 190 species of bees, from bumblebees to leafcutters, carpenters, and cactus-lovers, but no members of the *Apis mellifera* family. This brings its own problems.

In 1993, DNA analysis confirmed just how small the gene pool is in the States by tracing the family tree of colonies used by 22 commercial bee breeders in the southeast. Of the 142 colonies that were tested, just 4% were found to be maternal descendants of the black bee. This was the first honeybee taken to the US by the settlers but is no longer used commercially. The other 96% of colonies could trace their roots to Italy and the Balkans, from where races were imported in the 19th century.

According to Stephen Sheppard, entomology professor at Washington State University, eight races are known to have been introduced to North America from Europe. A ninth arrival was the descendants of African honeybees that were brought to Brazil in 1957 and spread northwards. But Europe's favourite bees dominate the North American landscape. Like their European counterparts, bee breeders in the US and Canada mainly sell strains of the gentle Italian, Balkan and Turkish bees, all of which make lots of honey.

Commercial beekeepers replace the queen in their colonies every one to two years to maintain good egg production. They purchase the queens from commercial bee breeders who rear them all from a relatively few selected breeder queens. The daughters are then mated with drones from colonies located in the commercial apiary. In 1993 the 22 apiaries in the DNA study used just 308 breeder queens to produce almost half a million new queens for sale. So when the study was conducted, about one sixth of all US colonies came from these 308 breeder queens. This creates the potential for a massive reduction of genetic variability.

In addition to the trio of dominant strains, three hybrids selected by bee breeders are popular with hobby beekeepers in the US: the Buckfast Bee, selected from crosses of various races bred in Britain; Midnite, derived from Caucasian bees, and Starline, from Italian stock. Midnite was developed to have an extremely gentle nature for the hobbyist. However, its honey-producing capabilities have also attracted the attention of commercial beekeepers.

The Buckfast Bee, which today is marketed in the US as an excellent choice for the east coast and northern states as it is well adapted to damp, cold winters, was originally bred by Brother Adam at Buckfast Abbey in Devon. He wanted to develop a bee with high resistance to tracheal mites after these parasites had allegedly wiped out thousands of colonies across the British Isles at the beginning of the 20th century.

Monasteries have a long tradition of keeping bees, going back to the middle ages when beeswax was turned into candles. When Brother Adam was put in charge of the Buckfast apiary in 1919, he set about rebuilding the 46 colonies that had perished. His intention was to use cross-breeding to create a gentle-natured, productive bee with low swarming tendencies and that was unlikely to die during winter. He travelled more than 100,000 miles in search of good traits in indigenous races that he could bring back to the abbey and incorporate into his bees. A breeding station is still maintained by the Abbey on Dartmoor.

But Brother Adam's early attempts at selection were not without their critics. Beowulf Cooper, an entomologist employed by the Ministry of Agriculture, raised objections to hybridisation, rejecting Brother Adam's pronouncement that Britain's native black bee had been wiped out by the Isle of Wight disease. In 1964, Beo, as he was known by his friends, formed probably the first honeybee conservation programme. The humble Village Bee Breeders Association is now the international Bee Improvement and Bee Breeders Association.

Although there were no native honeybees in the United States, feral populations soon built up as swarms made their homes in tree hollows, caves and the eves of derelict buildings. These unmanaged colonies were subjected to selective pressures that their managed sisters were not, and studies show that there were significant genetic differences between feral and managed colonies. A 1994 study, led by US Department of Agriculture research entomologist Nathan Schiff, which did DNA tests on 692 feral colonies in the southern US, found that 37% had genes associated with the original black bee brought over from northern Europe, compared with 4% in the commercial stock.

The significant genetic differences between commercial and feral populations made the latter a source of genetic variation for breeding programmes. The DNA studies concluded that the potential to select for desirable traits such as higher tolerance to pathogens and parasites could be enhanced in managed bees by including feral colonies in the breeding process.

Unfortunately, feral colonies in the US – like Europe's wild honeybee population 10 years earlier – have been pretty much wiped out by the varroa mite since its arrival in Florida in 1987.

Benjamin Oldroyd, a bee geneticist at the University of Sydney, says worker bees have numerous defences against disease, including behaviours in which some workers seek out diseased brood and remove it from the colony. To be effective, however, behavioural defences require a high level of genetic variation within colonies. "If all workers are the same, they may solve one problem brilliantly but be more vulnerable to others," explains Oldroyd.

In an essay called "What's Killing American Honey Bees?", he puts forward a little talked-about theory to explain colony collapse disorder that boils down to bees' defective gene pool.

Honeybees need to incubate their brood at a high temperature of 35C (95F). If the temperature of the hive's nursery drops a few degrees, the adult bees that emerge are normal physically, but studies have shown that they display deficiencies in learning and memory, says Oldroyd. Workers reared at lower temperatures tend to get lost in the field and can't perform communication dances effectively. This, he argues, means that if colonies are unable to maintain optimal brood temperatures, symptoms will occur that are similar to those reported for CCD: adult bees disappear, leaving just the queen and a few younger bees in the hive. "If any factors change the ratio of adult bees to brood so that the adults are unable to sufficiently regulate the brood nest temperature, then CCD-like symptoms will occur," he says.

Oldroyd also expects cool-brood bees to show higher levels of stress-related viral infections, leading to a vicious circle in the hive. "These effects could act synergistically – more viruses lead to shorter-lived, less efficient workers; that in turn leads to suboptimal temperature and more short-lived bees," he says.

But why would honeybees suddenly become less able to maintain a cosy home for their babies? Possibly lack of genetic diversity, says Oldroyd. Research conducted with colleagues at the University of Sydney on the thermoregulation of honeybee nests found that the more genetically diverse a colony of honeybees, the more able it was to maintain a stable temperature for its brood.

Workers regulate temperature in the hive by fanning hot air out when they think it is too hot, and clustering together and shivering to generate metabolic heat when they think it's cooler than the temperature required to incubate their brood.

Exposing two genetically diverse and two genetically uniform hives of bees to the Sydney winter showed that the temperature varied much less in hives containing genetically diverse bees. When the temperature was raised 5C (9F) above the optimum level, again the diverse colonies had a more stable response.

A number of other studies lend weight to the argument that genetically diverse honeybees are better at warding off disease. David Tarpy, an entomologist at North Carolina State University, demonstrated that when colonies were infected with the virulent American foulbrood disease, a bacterium that shrivels the capped larva and kills it, increased genetic diversity among the workers reduced the intensity of the attack. Genetically diverse colonies were also less likely to suffer from a fungal infection called chalkbrood.

A long-term comparison of the development of swarms of bees highlighted significant differences between those with many patrilines, and those with workers that shared only one father. The research, by Heather Mattila and Thomas Seeley of Cornell University's neurobiology and behaviour department, found that within two weeks of establishing colonies in a new nest site, the genetically varied swarms produced 30% more wax comb, stored 39% more food, and had foraging levels that were up to 78% higher.

When nectar was abundantly available during the summer and autumn, the nests inhabited by multi-father colonies gained more weight (mostly in stored honey reserves) than the homogenous bees. The researchers said this difference implies that a key advantage of genetic diversity may be related to how honeybees utilise sophisticated mechanisms for recruiting nestmates to forage, which could include communicating the location of food sources through waggle dancing.

By the end of the summer, the genetically diverse colonies were spectacularly healthier and stronger. Their populations were five times larger, they reared eight times more reproductive males, and their hives were significantly heavier, mostly because of large amounts of stored food. These surplus food reserves sustained them throughout the winter months, whereas all of the one-father colonies in the study starved to death by December. "Early disparities between colonies therefore translated into long-term effects on colony survival and productivity," the study concluded.

Mattila and Seeley said their results suggest that the use of poorly mated queens in commercial operations (mating quality includes the number of mates, the sperm stored, and genetic diversity of mates) may have a large impact on the colonies that they head.

This lack of genetic diversity is not confined to honey-bees: it is a feature of modern agriculture. As farming systems become more market-oriented, selection goals related to survival shift to prolificacy, early maturity, and high growth rate for anything from cows and chickens to rice and apples. In the case of dairy cows, high milk yield was the goal. Today the average US dairy cow produces twice as much milk as she did in 1957. This improvement results from genetic selection for milk yield, as well as improved feeding practices and other management changes. In the process fertility has declined. Lack of selection for reproduction, the stress of lactation, and changes in the cows' environment are all suspects.

Similarly, honeybees, which are described as livestock by commercial migratory beekeepers, have been selected for high honey yields, colony growth early in the year and gentle behaviour. Could high mortality be the price that has been paid?

Beekeepers in the US suffering huge losses from CCD turn to breeders in California and Texas to restock their hives. These commercial breeders are described as the backbone of the nation's bee industry. But if they are producing bees that are increasingly less genetically diverse, research suggests that the replacement colonies will fare no better than their predecessors.

Some commercial beekeepers have looked to Australia for help. Pennsylvania-based Dave Hackenberg, whose bees were reportedly the first victims of CCD, lost two-thirds of his 3,000 colonies in 2006-7. He bought 384 packages of bees – each containing a queen and three to four pounds of workers – from Australia to help restock his empty hives. But the antipodean imports fell at the first hurdle. Less than 10% survived the winter. "They were useless," says Hackenburg. "Only about 25 colonies are alive today."

Susan Cobey, who heads a honeybee biology research centre at the University of California, Davis, is not surprised that so many Australian bees perished. "They don't have a lot of resistance to varroa because they haven't been exposed," she points out.

In Britain, where varroa struck in 1992, replacement honeybees were often imported from New Zealand. Norman Carreck, who was a young apiculturist at the UK's leading bee research establishment, Rothamsted research station, at the time, remembers that the results were equally disastrous. Not only had varroa yet to reach New Zealand but Carreck recalls that the bees' body clock was preset for the southern hemisphere and they failed to adapt to the British seasons.

"When they should have been storing honey for our winter and the queen should have stopped laying, she was thinking it was spring and laying loads of eggs. So you'd end up with huge colonies going into winter with not enough food to keep them going and they'd starve," he says.

Cobey is working with Steve Sheppard, of Washington State University, to import bee semen and eggs from Europe to revitalise the US honeybee gene pool. "What we're proposing to do," she says, "is to bring in some semen and eggs from different subspecies and then do artificial insemination and breed from these. Basically, we want to bring some new genes into the gene pool that will be available for queen breeders to use."

However, a decades-old ban prohibits the importation of European bee stock. The 1922 Honeybee Act, which was prompted by huge losses of colonies in Britain believed to have been caused by the tracheal mite, was imposed to prevent the entry of honeybee diseases and parasites, as well as undesirable subspecies. It also restricts the importation of genetic material. However, Sheppard and Cobey are optimistic that they can convince the US Department of Agriculture's (USDA's) animal and plant health inspection service to lift the ban. Cobey points out that many of the diseases it was intended to thwart have long since made it into the States, including the tracheal mite.

In 2004, the ban was modified to allow the importation of bees from varroa-free Australia because of a feared shortage of bees to pollinate California's almond orchards. A decade earlier, USDA scientist Thomas Rinderer was allowed to bring in Russian bees under permit for breeding purposes. Cobey got a permit in 2006 to import bee semen from Germany, but when CCD was first reported later that year restrictions were tightened again.

Bret Adee of Adee Honey Farms, the largest commercial operator in the US, was looking to replace the 28,000 colonies he lost in 2008 with bees bred for varroa resistance from Rinderer's Russian stock. "The problem is," Adee says, "they are good for cold climates and take less maintenance but they're not a great bee for pollination or honey."

Cobey established the Vaca Valley Apiaries in Vacaville, northern California, in 1982. Here, she developed the highly regarded New World Carniolan breeding programme, selecting queens for a variety of traits including productivity, overwintering ability, resistance to tracheal mites, docility and high brood viability. "Breeding them [honeybees] so they're strong and healthy and resilient, so they will bounce right back, it's a passion and an increasing challenge," says Cobey. With Sheppard, she wants to bring in semen from the super-honey-makers – those Italian, Balkan and Turkish bees whose genes already dominate the US's ever-diminishing gene pool. They say it will at least expand the gene pool of these three races and increase the number of breeder queens that colonies are raised from.

"Many of the Italian and Carniolan [*carnica*] bees we have here are pretty mongrelised," says Cobey. "We're working with bee institutes in Germany, Turkey and Italy that have stocks that have shown some varroa resistance."

Asked why they don't bring in black bees, she responds that they are "a lot more defensive" – in other words, they sting more readily.

"It's a balancing act between the needs of the industry and healthy bees. The Italian bees produce brood all year round now but it is also mite food. Australian bees can be brought in just for the almond pollination, but then they crash because they are more susceptible to varroa."

She says the US will have to change its attitude toward bees and production. She believes the answer could be using Africanised bees, and cites Puerto Rico as a country where the Africanised bees are "more gentle but have amazing resistance to varroa".

Genetic selection should make this possible. If scientists can correlate genes to certain behaviour, in theory they could take the resistance genes from the Africanised bees and mix them with the gentleness and honey-making genes of the Italian bee.

But is genetic diversity the magic bullet in the fight against CCD?

Oldroyd thinks it unlikely. "I doubt that a narrow genetic base is the cause of CCD," he says. "I included it in my paper last year for completeness of all the theories at that time."

Australia has a similarly small honeybee gene pool to the US. With no indigenous races, it was not until 1822 that eight colonies of black bees arrived from Britain. Almost 200 years later, there are around 600,000 hives and more than 10,000 beekeepers. As in the US and Europe, only a fraction of Australian beekeepers are commercial operators owning 250 hives or more. But these 700 keepers (7% of the total) account for 70% of the hives.

The Australian Honey Council advises that commercial

beekeepers must produce more than 200-250kg (440-550lb) of honey per hive each year to remain viable. This compares with an average yield per hive of 67kg (148lb). With these pressures, it is no surprise that the three main races of honeybees kept on the continent are identical to those in the US and Europe. The main difference between the populations is that Australia's doesn't have varroa mites sucking its blood.

Most commercial beekeepers in Australia are migratory for honey. The climate allows for beekeeping for 12 months of the year in some states, and beekeepers move their hives from honey flow to honey flow as many as six times a year. Journeys of more than 1,000km (625 miles) are not uncommon. Many are increasingly carrying out paid pollination, mainly of apples, pears, almonds, melons, watermelons, pumpkins, rapeseed oil and seed crops, as more and more single crops, such as almonds, are grown over vast expanses of land.

Queen breeding is carried out by a small number of breeders. A 2007 survey of the genetic make-up of commercial and feral honeybees in Western Australia found similar results to the US. The commercial populations were derived from the Italian honeybee. But, surprisingly, the feral population was descended from both Italian stock and the *Apis mellifera iberica* race that originated in Spain. It is possible that these arrived via Spanish missionaries or imports from the US. No genes of the black bee were detected.

Given the close similarities between the genetic make-up of commercial honeybees in the States and Australia, and the stresses they both face, from increasing monoculture and droughts to long-haul journeys, you would expect antipodean bees to be collapsing. But there is no talk within Australia of CCD.

Oldroyd points out that there are often reports across the country of vanishing bees, a phenomenon that Australians call "disappearing disease". He questions whether CCD is anything new – or just part of the continuum of bee losses.

CHAPTER 4

ONE CRISIS AFTER ANOTHER

It was a fine morning when Jesse Newson walked down to his small, sheltered orchard to open his hives. He carefully lifted the roof off the first and was concerned to see only a handful of bees at home. He opened the next to find even fewer, while the third had none at all. Never, in his 25 years of beekeeping, had Newson seen anything like it. He racked his brain for a reason. Was the honey flow so good that more bees than normal were out collecting the nectar? Had they somehow lost their sense of direction? Perhaps they had been delayed and would be home soon? He knew all this was wishful thinking. Bees are methodical creatures with a sophisticated navigation system. They usually have babies to feed, a colony to support with nectar and pollen. They just don't go awol. But he waited and waited in the hope that they would return in a few days. They didn't. Closer inspection showed there were no larvae waiting to be fed, but the strangest observation was that the hives were full of delicious honey. Try as he might, Newson could not work out where his bees had gone, why they had gone, and why they hadn't eaten their food stores beforehand.

At the end of the beekeeping season, he wrote: "I have not seen so disastrous results among bees as in the present year. I have lost 19 stands [hives] since first of November; in some of them as many as 40lb of honey were left looking very nice and tasting as well as any I ever saw."

Newson's experience in Bartholomew County, Indiana, was not unique in that part of the United States. In nearby Scott County, apiarist TJ Connett reported, "There is a disease prevailing to an alarming extent among our bees this fall that is entirely new, nobody being able to find any cause or remedy. Old and substantial swarms die, leaving the hive full of honey."

And another Indiana beekeeper reported that not only had he lost 40 hives, but all the bees as far as 30 miles north and 18 south had "stampeded in the same manner as mine, to the hive of mother-earth".

In neighbouring Kentucky, JN Webb found a similar ailment sweeping away his apiaries. "Many swarms left well-stored stands of excellent honey, amply sufficient to carry them through the winter; and what is more strange, comparatively few of the bees were found dead in the hives," he recorded.

He could have spoken for the whole beekeeping community when he added: "What was the cause of the wholesale destruction of this useful and interesting insect, dying in the midst of plenty, away from the hive, we cannot understand."

These accounts of a strange plague wiping out bees could be describing colony collapse disorder (CCD), the name that has been given to the current unexplained disappearance of honeybees across the globe. As with CCD, the bees seemed to vanish without trace, leaving large honey stores behind. Yet the observations by Newson and his peers refer to a disease that swept through Indiana, Kentucky and Tennessee almost 140 years ago.

Ever since this first published record of bee disappearances in 1869, it seems that hardly a decade has gone by without a mysterious large-scale die-off of *Apis mellifera* somewhere in the world. And with each strange malady comes a new name. In 1891 and 1896 an outbreak of what became known as "May disease" – because it occurred in that month – was recorded in Colorado. In 1915 it was New Jersey, New York, Ohio and Ontario's turn to experience massive honeybee losses, and the northern US was hit again in the 1930s with spring or winter "dwindlings". In the "fall dwindling" of the autumn and winter of 1963-4 several thousand colonies, stretching from Louisiana right across to California, died of unknown causes. A Louisiana beekeeper with 55 years' experience said he had never seen anything like it.

In the 1970s the term "disappearing disease" was coined for a puzzling phenomenon that wiped out bees in 27 states. But unusually high bee mortality was not confined to America. Australia is thought to have witnessed a wholesale disappearance of bees as far back as 1872. In 1910, R Beuhne, president of the Victorian Apiarists Association, estimated that beekeepers around the town of Stawell had lost 80% of their bees. Sixty years later, Ken Olley, a commercial beekeeper based in Brisbane, raised the alarm about a decade-long die-off in commercially run apiaries throughout the country. And in France, as recently as 1995, apiarists warned of their bees "melting away" in massive numbers.

Until the current crisis, Britain was home to probably the most famous bee epidemic. First spotted on the Isle of Wight at the beginning of the 20th century, in five years it had annihilated 90% of the island's honeybee colonies. By 1918 it was reported that "not a beekeeping district in Great Britain was free from the scourge". What was unusual about "Isle of Wight disease" was that the afflicted bees did not abscond, but were seen crawling out of the hive in their thousands, unable to fly. Dr John Anderson, who studied the strange spectacle, noted how the bees would "crawl out of the hive on a fine day and litter the ground in a fan-shaped space extending many yards in front of the hive. In this space one could not put a foot down without crushing at least one bee."

Almost 10 years after beekeeper HM Cooper discovered the disease, he despaired that no matter what preventative or curative measures he tried, it "has triumphed over all, and so far as our present knowledge goes, we are in a hopeless case".

His words are echoed today by beekeepers up and down America, as they experience a second year of colony collapse disorder. "We're still no nearer knowing what is the cause or how we can stop it," says Dave Hackenberg, the Pennsylvania beekeeper who is credited with discovering the latest plague in the autumn of 2006.

In the past, as now, speculation about the causes of vanishing bees was always far more plentiful than concrete evidence. In 1869, the finger was pointed at a lack of pollen, poisonous honey and an unusually hot summer. But by 1917, an overabundance of pollen in the bees' diet was blamed for their demise. "When the affected bees were bust, the abdominal contents were found watery and gave off an unpleasant, somewhat pungent odour," Elmer Carr reported in his account of an investigation into the death of thousands of bees near Bradevelt conducted by New Jersey's Apiary Inspection Department and a representative of the Bee Culture Laboratory in Washington.

In Australia, honey from eucalyptus trees carried the can for the bees' alarming death rate. It was found to ferment too quickly during the winter, and as a result the bees starved.

Researchers at Cambridge University were certain they had cracked Isle of Wight disease when, in 1912, they found the parasite *Nosema apis* in the gut of the infected bees. But a few years later Dr John Rennie at Aberdeen University identified another blood-sucking parasite, a mite named *Acarapis woodi*, which lives in the bee's trachea, or windpipe, making it difficult for the insect to breathe. Since the Isle of Wight symptoms were found in bees that were heavily infested with *Acarapis woodi*, the affliction was renamed acarine disease. The only problem was that the mite was also found in bees that could still fly.

Lesley Bailey, a British parasitologist, decided to try to get to the bottom of the disease in the 1950s when he joined the UK's foremost agricultural research establishment, Rothamsted experimental station. He knew that something was wrong with the official version of events when his attempts to reproduce the symptoms associated with acarine disease failed in honeybee colonies infested with the tracheal mite.

Bailey said it was easy to understand why the mite had become a scapegoat. It was large, and the incidence and appearance of bees' infested tracheae was "startling", he wrote. "Lacking knowledge, yet feeling the need for an explanation, they assumed an infectious disease was killing the bees."

With similar symptoms turning up in France, Switzerland and Russia, it was a natural assumption to make. Yet, Bailey points out: "This is almost remarkable progress if it started in England, especially considering the difficulties there must have been transporting bees during World War I and the Russian Revolution." He accused the British press of printing a number of "sensational but uninformative" articles about catastrophic bee blight that firmly rooted the idea of an infectious disease in the collective consciousness.

In his lab, using one of the first ever electron microscopes, Bailey identified the Isle of Wight killer as chronic paralysis virus, or CPV, which caused bees to lose the ability to fly. This was not, he insisted, a new disease. Bailey was convinced that bees and parasites such as mites had evolved together for several million years. Only when the bees' immunity system broke down could the mites spread the infection that would overwhelm and destroy the honeybee colony, by either activating or promulgating a normally latent virus.

So what weakened the bees' immune system to let CPV take hold in the early 20th century? Poor environmental conditions, said Bailey, such as an unseasonable spell of wet or cold weather that would leave the bees hungry and confined to the hive.

The spring of 1906 – the worst year for bee losses on the Isle of Wight – is described as being "very early" and "hot enough to draw crowds to the beach", followed by frost and snow in April. Bees that had spent March foraging for nectar and pollen were suddenly shut up in their hives and could have become very dysenteric and run down, enabling the spread of the tracheal mite and the virus, said Bailey.

Australia's more recent bee disappearance was thought to be aggravated by damp conditions and poor nutrition. Honeybee losses in the Rio Grande region of Texas in the 1970s occurred after a period of unseasonable cold followed by two weeks of rain. And a particularly long, cold winter during 1995-6, followed by a wet spring in the north-east US, preceded a region-wide epidemic.

Yet the link between bees' health and weather conditions is nothing new. Aristotle noted that bees become sick "especially when the blossom on the trees is mildewed, and in years of drought".

Annual reports from the British Beekeepers Association indicate that the so-called Isle of Wight disease was on the wane from 1913 to 1926. And by 1929, Anderson was able to report: "Such a very heavy death rate is now quite rare and can be seen only when conditions are extremely unfavourable to the bees."

Mortality rates during the early 20th century were accelerated by beekeepers feeding both ailing and healthy bees a potpourri of poisonous potions. Sour milk, salt and beef extract were some of their favourites, administered in a syrupy solution. One of the most ruinous practices was stripping the hive of pollen, too much of which was thought to extend the bees' recta, and feeding the bees these "remedies" instead. A bloodcurdling treatment of petrol and benzene was created by a Lincolnshire beekeeper, Richard Frow. Even in the 1970s, by which time you might expect beekeepers to know better, bees were still being fed harmful pollen substitutes. An investigation into a mysterious disappearance of honeybee colonies in Florida, by Ohio's department of entomology and the US Department of Agriculture's bee breeding and stock centre laboratory, found that bees fed soya bean flour were most likely to perish.

In 1975, an extensive survey of beekeepers across the US indicated that bees had vanished in a spectacular fashion in as many as 27 states over the past 15 years. Nine of the 46 inspectors who responded to the survey estimated that more than 43,000 colonies had been wiped out in five years. Fifteen of them blamed pesticides for the losses.

Beekeepers were no strangers to the dangers of modern pesticides. Since the 1950s, highly toxic chemicals had been liberally sprayed on crops throughout the United States. The chemical compounds were designed to eradicate a variety of crop destroyers. The problem was that they often killed the pests' natural predators, too, as well as other beneficial insects.

In 1953, a New York State beekeeper reported losing 800 colonies after the state sprayed a large area of orchards with DDT. So widespread and heavy was the loss that year, according to Rachel Carson's famous book Silent Spring, that 14 other beekeepers joined him in suing the state for $250,000 in damages. Another beekeeper, whose 400 colonies were incidental victims of the US government's all-out chemical war on the gypsy moth in 1957, reported that all the foraging bees had been killed in forested areas and up to half destroyed in farming areas that were sprayed less intensively. "It is a very distressing thing," he wrote, "to walk into a yard in May and not hear a bee buzz."

The following year, it was the fire ant's turn to have the chemical guns turned on it when a million acres (400,000 hectares) of agricultural land in the southern states were sprayed with dieldrin and heptachlor, two relatively new insecticides that were many times more toxic than DDT.

Excessive bee kills were commonplace in the latter part of the 1960s. Arizona, which along with Washington, California and Georgia, received the most federal money to compensate beekeepers, lost almost half its bees between 1963 and 1977. When farmers were no longer allowed to use DDT because it was considered too toxic, they switched to other pesticides. The most lethal was Penncap-M, whose tiny capsules, the size of pollen grains, were collected by bees and carried to hives.

Yet despite many examples of the damage caused by pesticides, the authors of the 1975 bee survey, William Wilson and Diana Menapace, both from the USDA's honey-bee pesticides and diseases research laboratory in Wyoming, ruled this out as the cause of vanishing bees.

They said there was no evidence of large-scale spraying, or of chemical residues found in plants, in any of the areas reporting the disease. They also claimed that bees poisoned by pesticides behaved very differently from those suffering from the "disappearing disease". With pesticide poisoning, foragers died in the field and could be seen lying on the ground, or nurse bees died crawling and tumbling from the hive entrance. With disappearing disease, bees dwindled in number during a period of inclement weather, as the worker bees disappeared from the hive to leave just a handful of bees and the queen.

The four main characteristics of disappearing disease, identified at a conference convened by a concerned American Honey Producers Association, were similar to those identified by Wilson and Menapace: colonies die, or dwindle, with plenty of honey stores; the queen is usually still alive; most of the bees usually die in the field; and most cases occur in the spring or autumn, when the weather is cool. Poisonous pollen or nectar, and diseases caused by micro-organisms, including viruses, were all ruled out as causes. But researchers failed to come up with any convincing answers because they were having problems reproducing the disappearing disease in laboratory conditions.

One hypothesis that gained prominence at this time was that bees had inherited some sort of genetic disorder through excessive inbreeding. Wilson and Menapace pointed to "genetic deficiencies that are enhanced by stress (ie, cool, damp weather)". Commercial beekeeper Charles Mraz, best known for founding the American Apitherapy Society to promote the therapeutic uses of bee venom, visited one of the best honey-producing regions of Mexico in 1977 following reports of disappearing bees. He advised the local beekeepers that "the only cure was to eliminate the bad stock with vigorous resistant queens". But he warned that they wouldn't be easy to find. "In both the US and Mexico intensive inbreeding over many years has pretty much eliminated the old, hardy stock we had years ago."

PF Thurber, a former Washington state apiary inspector, blamed genetic flaws for making his bees fall from the sky. "There were clouds of bees," he wrote. "Unbelieving, I walked out in the pasture and was suddenly aware bees were falling like rain. They flew out, chilled, fell and died, or died and fell. Tears ran down my cheeks because I could do nothing."

Thurber sought help from a specialist bee disease establishment, which informed him his bees had a "genetic defect". On inspection of the two hives where he raised queens for his apiary, Thurber found that one had in front of it a "fan of dead bees over four inches deep ... There must have been 15lb of dead bees."

Thurber's description bears an uncanny resemblance to Anderson's some 50 years earlier in Britain, but no link was made to chronic paralysis virus. He advised fellow apiarists to "cross your fingers and have faith" – advice that many beekeepers are still following today.

There is little doubt that the honeybee gene pool across the United States was small to start with because, unlike in Europe, there were no native honeybees with which imported varieties could breed. And it shrank still further when the Americans banned the importation of foreign stock. The 1922 Honeybee Act was fuelled by fear that Isle of Wight disease would spread to the US. By the 1970s, the supply of new queens in the US was increasingly in the hands of a few commercial breeders.

In 1977 and 1978 Ohio State University entomology department and the USDA's bee breeding and stock centre laboratory set out to test if genetic weakness was behind the bees' demise, comparing bees that had "disappearing disease" with colonies in the same apiaries that displayed no symptoms. But the researchers found no evidence to support the genetic hypothesis. Instead, they attributed the die-off to a range of "environmental factors", primarily a lack of nutritious food, because of a cold or wet spell of weather, coupled with poor beekeeping practices.

Honeybees, like all animals, are susceptible to a number of pests and illnesses. If not kept in check, these can drastically reduce a colony's birth rate and average lifespan, throwing into jeopardy the delicate cycle that ensures its survival. The riddle of the vanishing bees can never be solved without a full understanding of the role played by these diseases and pests, some of which have afflicted honeybees for millions of years.

As long ago as 350BC, Aristotle gave an account of bee maladies in his History of Animals. "Small creatures are produced in the colonies, which damage the honeycombs," he wrote. "One is the little grub that makes a web and is damaging to the honeycombs (it gives birth within the honeycomb to a sort of spider resembling itself, and causes disease in the colony); another is a creature like the moth that flies round the candle; this gives birth there to some-thing full of down and it is not stung by the bees but flies away only when smoked out." The first insect Aristotle was referring to is probably a beetle called Trichodes (Clerus) apiarius; the second, the wax moth.

He also described a disease that made the bees listless

and produced a bad smell in the hives. This is now known as foulbrood, of which two equally lethal versions exist: European and American.

Foulbrood, as its name implies, goes for baby bees. It is caused by bacteria that consume the larvae and are spread around the hive by adult bees trying to clean their home to get rid of them. Foulbrood seems to have been a problem for US beekeepers ever since the Pilgrim Fathers set foot in the New World. During the 17th century it may have been responsible for wiping out colonies along the Atlantic coast; the only cure was to burn both hives and bees.

Chalkbrood is another baby killer, a fungus whose victims die shrunken and hard. Their chalk-white remains can easily be seen on the landing board at the hive's entrance as the house bees try to clean out the dead larvae. A strong colony can live with a small amount of the fungus, but it can destroy a weaker one.

The larvae of the small hive beetle, meanwhile, feed on the bee's brood, pollen and honey. Unlike the wax moth, which can be chased out of the hive by a strong colony's own police force, the beetle – originally from South Africa – will destroy the hive unless the beekeeper stops it.

Nosema apis, a parasite that lives in the gut of an adult western honeybee, causes dysentery and weakens the bees, reducing their life expectancy and depleting the colony. It is now treatable with antibiotics, but its closely related cousin, *Nosema cerana*, which used to live only on the Asian honeybee, has now spread to the west, causing similar symptoms that are being treated with the same antibiotic with varying degrees of success.

The tracheal mite, *Acarapis woodi*, also causes premature death, usually in the winter. Without enough adult bees to feed them in the spring, many of the expanding brood also die.

America's attempts to keep out many of these life-threatening diseases and pests failed. In 1972 chalkbrood appeared among beehives in California. It was joined by the tracheal mite in 1984, and the small hive beetle (which has yet to reach Britain) followed a few years ago, wreaking havoc across the south-eastern states. It was only a matter of time before the deadliest of all parasites would arrive fresh from massacring Europe's honeybee population.

The varroa mite had happily co-existed for centuries with *Apis cerana*, the Asian honeybee. But when the bee's western cousin was taken to Asia early in the 20th century, it soon became apparent that *Apis mellifera* and the varroa mite weren't going to get on so well. In fact, the parasite developed a new species specially designed to attack the western honeybee, known as *Varroa destructor*. Like something out of a horror movie, these blood-sucking mites hitch a ride into hives on the backs of unsuspecting bees. Once inside, the female mites bury themselves at the bottom of the brood cells, feed on the larvae and lay their eggs, which hatch and mate and continue the cycle until the bees are wiped out.

The bees that mites have fed on have a shorter lifespan, as well as shrunken and deformed wings, and are less resistant to infection. It was initially thought that the mite's feeding action made the bees' bodies shrivel. But studies revealed that the deformity was caused by a latent virus called deformed wing virus that was triggered by the mites. Further studies discovered more problems in heavily infested bees such as acute paralysis virus (APV), a faster-acting version of the chronic paralysis virus (CPV) that was identified years earlier as the Isle of Wight killer.

When varroa spread to Britain in 1992, Norman Carreck was a young entomologist at the Rothamsted experimental station where Bailey had made his name. We arrange to meet Carreck at his home in the Sussex countryside on a cold, wet morning in January. He collects us at the station in his old banger and drives us along winding country lanes until we reach the large, ramshackle house he shares with his elderly mother.

He ushers us into his study crammed full of books about bees. Two years ago, Carreck was a victim of government cuts to bee research at Rothamsted, where he worked for 15 years after graduating in agricultural studies from

Nottingham University. He is now senior editor of the international Journal of Apicultural Research, and sits on various beekeeping and research association boards.

Over cups of tea and biscuits in front of a log fire, he shares with us his knowledge of the devastating role varroa has played in the honeybee's history, and his theories for what lies behind its latest troubles.

Carreck recalls that the year after he joined Rothamsted, he and his colleagues went to Devon, where varroa had first been identified in Britain, to collect samples of the mite-infested bees for analysis. They found yet another disease, he says: slow paralysis virus (SPV), which allows the bees to survive for a few more days than those with APV.

Carreck says he personally witnessed whole colonies walking out of a hive. "It was a very curious event," he recalls. "It was late summer and the colony appeared big and strong. I'd looked at it and decided it was absolutely fine: I didn't need to look at it again. But when I went back a few days later to put mouse guards on the hive in preparation for winter, I saw several large colonies in the process of walking out of their hives. They were New Zealand [western honey] bees so they were very pale and I could see them clearly against the grass, all crawling away. If I'd gone a few days later, I would have found an empty hive, maybe with just a queen wandering around."

Studies have not been able to determine whether the

varroa mite acts as a vector to spread disease pathogens and impose a stress that aggravates diseases such as nosema, or if it somehow activates usually latent viruses to multiply. But there is no doubt that its large-scale presence has spelled death to bee colonies worldwide.

Prior to varroa's arrival in the US in 1987, average winter honeybee losses of around 10% were recorded in Pennsylvania. In 1995-6, however, thousands of colonies died in a region-wide epidemic in the north-eastern United States. Pennsylvania's beekeepers reported average colony mortality of 53%, while in Maine as many as 80% of bees were wiped out. Although some of the losses were attributable to a "particularly long, cold winter, followed by a wet spring", scientists investigating the cause of the killer said "weather alone cannot be blamed for losses this severe".

By this time a number of varroacides had been licensed. The researchers found that bees that weren't being treated with these chemical compounds, or others to control nosema, had significantly higher mortality rates than those that were. This led them to conclude: "Aggressive treatment for honeybee mites and other diseases significantly increases colony survival."

Moreover, they foresaw continued colony losses in the years to come unless new methods of control were developed. And they called for further research to "determine precisely how parasitic mites and diseases associated with them have a detrimental effect on colony survival" and to "understand how mite diseases interact with other honeybee pathogens".

More than 10 years on, as scientists grapple for the answers to perhaps the deadliest and certainly most costly bee epidemic ever to hit the planet, they still don't know the answers. Nor are they clear how mites might make bees more susceptible to exposure to pesticides or to GM crops, or conversely whether a substance in a pesticide or a GM crop could make the bee more susceptible to a pathogen.

This is of crucial importance because not all mass bee deaths in Europe in the past 30 years have been attributed to varroa. Pesticides are still a prime suspect because many of those used on bee-pollinated crops are toxic to honeybees, and GM crops have not been ruled out as an accomplice.

In 1981, use of Penncap-M was curtailed in the US after farmers who had sprayed the chemical were sued by Connecticut beekeepers for killing their bees. More recently, French beekeepers got their government to ban a leading brand of pesticide. BayerCrop Science, the manufacturer of imidacloprid, which was sold under the trade name Gaucho in France as a treatment for sunflower seeds, cites studies to prove its product wasn't responsible for the deaths of hundreds of thousands of colonies of bees in the mid-1990s. But Maurice May, spokesman for the 50,000-strong French beekeepers' association the Union Nationale d'Apiculteurs, said: "Since the first application of Gaucho we have had great losses in the harvest of sunflower honey. Since the agent stays in the soil up to three years, even untreated plants can contain a concentration that is lethal for bees."

Systemic pesticides like imidacloprid work their way up through the plant into the flowers' pollen and nectar and will be consumed by bees at low, but potentially dangerous, levels. Studies in France and Italy have found that imidacloprid can disorientate bees, impair their memory and communication, and cause nervous system disorders.

Pennsylvania state apiarist Dennis vanEngelsdorp, who is investigating the cause of colony collapse disorder in the US, says tests for pesticides have traditionally focused on finding doses that are lethal to adult honeybees. "At sublethal doses, the effects are more subtle. Honey production suffers and foragers seemingly disappear. For example, when colonies are placed near crops of sunflower treated with imidacloprid, foraging is disrupted and colonies dwindle and die as foragers fail to return to the hive. This may be explained by the fact that some pesticides can cause a disturbance in the dance language or in the orientation abilities of worker bees."

He says much more work needs to be done into the sublethal effects on both adult bees and larvae. Carreck agrees. He says the new neonicotinoid family of pesticides to which imidacloprid belongs are much less toxic than the organophosphates and chlorinated hydrocarbon pesticides that were developed after the second world war, but as a result we have probably become rather blasé about their effect on bees. "While not a danger on their own, what about when they get mixed with pollen or other chemical compounds?" he asks.

With neonicotinoids now found almost everywhere in North America and Europe from front lawns to apple orchards and fields full of oil seed rape, many experienced US beekeepers are siding with their French counterparts. But most scientists remain as unconvinced now as they were in the 1970s that pesticides are the culprit. "Why have bees disappeared from areas where no pesticides are used?" they ask.

Interestingly, the French are still losing bees on a massive scale despite the ban on Gaucho for use on sunflower seeds. A study of 41 apiaries from various regions of France, which in some cases reported 100% mortality rates in 1999-2000, found that in more than 76% of apiaries, one or several serious diseases known to be able to totally destroy colonies were present. Chronic paralysis virus emerged as a major cause and was dubbed "maladie noire" as the bees became hairless, black and sick. But the study hedged its bets, attributing the bee deaths to a possible variety of factors including latent diseases that the bees were too weak to fight when they became cold and hungry during winter, and too many colonies in single-crop fields, which prevented them from having a balanced and nutritious diet.

Like their American peers a few years earlier, the French researchers concluded: "Whatever may be their original main cause, winter losses are often accompanied by very aggressive pathogens whose impact cannot be neglected. Preventative measures and treatments must be applied."

Germany suffered its own wipe-out of bee colonies during the winter of 2002-3, and there are almost daily reports of colony collapse disorder killing bee populations from Portugal to Croatia and Greece.

Whether all these mysterious bee deaths can really be attributed to CCD remains to be seen. From the so-called Isle of Wight disease in 1905 through to disappearing disease in the States in the 1970s and now colony collapse disorder at the beginning of the 21st century, periodic bee calamities have often been lumped together when each outbreak could be due to entirely different factors. Wilson and Menapace were so sceptical of any connection between bee deaths they investigated that they recommended the name "disappearing disease" be dropped. The UK and Canadian governments have been keen to stress they don't yet have colony collapse disorder, despite some reports in these countries of inexplicable bee losses.

VanEngelsdorp sums up the predicament thus: "The losses that have been occurring for over 100 years could be completely separate events or part of a cycle of disappearance. So far we can only speculate."

The USDA's Jeff Pettis says the researchers he is coordinating have yet to find the "smoking gun". Israeli acute paralysis virus has been identified in 96% of hives affected by CCD, but Pettis admits that the virus could prove to be a symptom – like *Nosema apis* and tracheal mites before it – rather than the cause of the current vanishing plague.

Just as the disappearance and deaths of bees puzzled and perplexed scientists and beekeepers more than a century ago, they still baffle and bewilder their descendants. This catastrophe is proving to be as much of an enigma in 2008 as it was in 1869.

CHAPTER 5

COLONY COLLAPSE DISORDER

At 3pm on November 12 2006, commercial beekeeper Dave Hackenberg walked into his yard in Rushkin, Florida, lit his smoker to make his bees a little sleepy and started to give the hives a few puffs. After a few minutes he realised that no bees were leaving. Perplexed, he jerked the lids off. He was stunned by what he found inside: the bees had vanished.

"I got down on my hands and knees and crawled around. And there's no dead bees anywhere. They're gone," says Hackenberg. "There was nobody home. It was like somebody came from outer space and swept them away."

Since that fateful day, Hackenberg has endlessly repeated the story of how he discovered the mysterious malady, known as colony collapse disorder (CCD), that has wiped out more than a third of all honeybees in the United

States and possibly millions more across the globe. When we finally got to meet him more than a year later in California's almond orchards, it's as if we know him. We've seen his lanky physique on TV documentaries, heard his twangy vowels on radio shows and podcasts, and read his pithy quotes in countless articles.

All of the 400 colonies in Hackenberg's Florida yard had been healthy when he brought them south from Pennsylvania in autumn 2006. A few weeks later, only 40 survived. As he inspected the hives that November, he found most were empty of adult worker bees, who had left their queen and brood. Just as mysteriously, the honey stores had not been robbed by other bees, wax moths or hive beetles.

"It was just the weirdest thing," he says. "It doesn't work this way. Normally, other bees would have been in the dead beehive robbing honey out, and the moths and beetles would have got in quick – but it was like a ghost town."

In all of his 40-odd years beekeeping, Hackenberg had never seen anything like it. He gathered the few dead bees that were left and asked Pennsylvania's state apiarist to take a look. A state apiarist, like a regional bee inspector in the UK, is a government employee who inspects hives in case of disease and provides advice to beekeepers on disease recognition and control.

The first tests showed many abnormalities, including

scarring on the digestive tract, swollen kidneys and evidence of a fungal or yeast infection in the sting gland.

A month after the dramatic scene in Florida, researchers at Pennsylvania State University, along with both Pennsylvania's and Florida's departments of agriculture, issued a preliminary report on what they dubbed "fall dwindle disease". Based on interviews with seven commercial beekeepers from Florida, Georgia, North Carolina and Pennsylvania, the report revealed losses of colonies of between 30% and 90%. The research implied that common beekeeping practices, such as transporting bees over long distances and reusing brood comb from diseased hives, were to blame for weakening the bees' immune system and potentially spreading disease.

As a result, the disease was dismissed by many in the industry as PPB ("piss-poor beekeeping") or "Hackenberg's hoax". But by the spring of 2007, Hackenberg, who by that time had lost two-thirds of his 3,000 total colonies, had plenty of company in his misery. When beekeepers from all over the States began opening up their hives after the winter, as many as a quarter found many of their bees had gone.

The United States suddenly woke up to the disastrous effect that vanishing bees could have on food supplies and agriculture. With honeybees vital for the pollination of 90 crops worldwide, a lack of bees could lead to a shortage of fruits and vegetables, cattle feed, nuts, seeds, even cotton.

A study by Cornell University valued bees' pollination services as worth $15bn to the US economy.

The plague was renamed colony collapse disorder and a group of scientists, led by Jeff Pettis of the USDA agricultural research service, and Penn State University entomologist Diana Cox-Foster, started to work flat out to isolate the cause and find a remedy for what was fast shaping up as a national catastrophe. They had a number of leads to follow in their hunt for a killer that had now struck in 22 states.

By late spring CCD had made headlines around the world. The more fanciful theories for the hive collapses included an al-Qaida plot to wreck US agriculture, radiation from mobile phones and even celestial intervention in the form of a honeybee rapture.

The main suspects were a new or mutated virus; fungal infections that could produce toxins; environmental contaminants, such as pesticides applied in the field or used to control mites in hives; nutritional stress, perhaps linked to drought conditions; migratory stress from confining bees during long-distance transportation, and increasing contact among different colonies; or a combination of some, or all, of these factors that together was suppressing the bees' immune system and allowing pathogens to cause the final destruction. It wasn't long before the calamity was being referred to as bee Aids.

At the end of March 2007, CCD was the subject of a

congressional hearing, with testimonies from beekeepers, almond farmers and scientists. Richard Adee, of Adee Honey Farms, painted a scary scenario of thousands of unexplained die-offs across the country hitting beekeepers large and small, migrationary and stationary. In Ohio, all but 100 of a highly respected beekeeper's 800 colonies were decimated, a Missouri beekeeper reported that only 104 of his 700 colonies survived, and a Kansas apiarist had only 1,650 hives remaining out of 4,400.

"There is no discernible pattern to these losses," said Adee. "One beekeeper may experience pervasive colony collapse, while neighbouring beekeepers report no such losses." He warned that agriculture, along with mites and pests, was placing increasing demands on bees and beekeepers. His own 70,000 colonies, while still alive, were at only three-quarters of their usual strength. "Rather than growing substantially, as they always have done, our colonies in California seem to be declining," he told Congress. It was to prove a worrying sign.

May Berenbaum, the Illinois University entomology professor who chaired the National Research Council's committee on the status of pollinators in North America, highlighted the danger honeybees faced even before CCD. "If honeybee numbers continue to decline at the rates documented from 1989 to 1996, managed honeybees will cease to exist by 2035," she warned.

Cox-Foster told Congress that bees with CCD are "infected with an extremely high number of different disease organisms". While she said there was little evidence of blood-sucking parasites such as varroa or tracheal mites, she did raise concerns about fungal infections whose presence could indicate the bees' immune system was suppressed. The fungi they were seeing, she said, were possible contributors to CCD because they had been reported to afflict bees in the 1930s, when they were linked to toxin production.

Cox-Foster also highlighted concerns over a new breed of pesticides "known to be highly toxic to honeybees". "Some research has suggested that these systemic pesticides can translocate or move through plants to become localised in pollen and nectar at concentrations that may affect bees," she said. "It is essential to determine whether these pesticides play a role as a causal factor in the CCD symptoms."

While her colleagues in the fields of toxicology and environmental chemistry pursued the pesticide lead, Cox-Foster took the pathogen angle, calling on the services of a leading epidemiologist to help her try to find the culprit. W Ian Lipkin, a doctor and professor of epidemiology, neurology and pathology at Columbia University, was known for uncovering the West Nile virus in humans by using DNA sequencing methods. Cox-Foster made a persuasive case for applying his revolutionary, rapid genome-sequencing methods to diagnose viruses infecting bee colonies.

Scientists at the University of Illinois had fortuitously unravelled the genetic code of the honeybee the previous year, making it possible to do the sequencing and then eliminate the genetic material of the bee.

They mashed up healthy and infected honeybees collected over three years from across the States, then, using this new method called metagenomics, searched the mush for viruses, bacteria, fungi and parasites. They also tested royal jelly imported from China and healthy bees imported from Australia.

There were no surprises when the results identified a catalogue of micro-organisms that honeybees harbour, including seven viruses associated with disease. A worldwide survey of honeybees' health in 2005 had found few countries where bees weren't riddled with parasites, pests and viruses.

Israeli acute paralysis virus was diagnosed in all but one of the sick colonies. It was also in half of the royal jelly samples and in the Australian bees. IAPV is a close relative of the widespread Kashmir bee virus, which can kill honeybees in three to four days. KBV was first identified in the western honeybee in the 1970s at the UK's Rothamsted experimental research station, in lab tests conducted with the eastern honeybee *Apis cerana* sent from Kashmir. Symptoms included weakened colonies, and hairless, dying bees trembling outside the hive. Before the arrival of varroa, which triggers the virus, KBV was a non-virulent academic curiosity in Europe and the US.

IAPV was discovered in Israel in 2002, but had not been found in the States before. Its appearance was hailed as a breakthrough when the results were published in Science magazine in September 2007. Dr Lipkin described IAPV as a "significant marker" for CCD. But Pettis, who co-authored the report, was careful to say that the discovery did not mean IAPV was the sole cause, or even a cause. "What we have found is strictly a correlation of the appearance of IAPV and CCD together," he warned. "We have not proven a cause or an effect connection."

The results also caused a sensation because they implicated Australian bees as carriers of the deadly virus. Since 2004, the US had lifted its ban on imported honeybees, and live bees were a fast-growing export commodity from Australia to the United States – never more so than now, as beekeepers attempted to repopulate hives emptied by colony collapse disorder. In 2007, Australian beekeepers had sent more than 30,000 packages of bees to the United States, but there were fears that the USDA would now consider reintroducing a ban.

A spokesman for the Australian Honeybee Industry Association was quick to leap to his country's defence, claiming the report had unfairly jeopardised Australia's $5 million (£3.6 million) live bee export market. "We believe that it's unscientific. The jury is still out on what causes colony collapse disorder – or, for that matter, whether there is such a thing as colony collapse disorder," Stephen Ware told the Australian Broadcasting Corporation.

Fortunately for Ware and those he represented, follow-up screening of several hundred honeybees from 2002 let Australian bees off the hook, as it found that IAPV had been in the States for at least two years before the US started to import honeybees.

The worldwide bee-health survey had, however, warned about the international shipment of bees. "Through unregulated trade of bees and bee products and often a lacklustre concern for established regulations, humans have made cosmopolitan many bee diseases that were originally limited in distribution," wrote James Ellis and Pamela Munn. "As a result of this and the chemicals we use to treat colonies for various diseases, bee health in many countries has been severely compromised." They concluded that humans had done a "relatively poor job [of] limiting the spread of bee diseases and pests globally".

It was only a matter of time before reports of bee disappearances fitting CCD's description began filtering in from across the world. In Taiwan, the TVBS television station reported in April 2007 that around 10 million bees had gone awol, with beekeepers in three regions reporting heavy losses. One beekeeper on the north-east coast told the United Daily News that six million insects had vanished "for no reason", while another in the south said that 80 of his 200 hives had been emptied. While the exact reason for the exodus was unknown, experts said "volatile weather" might be to blame.

In Britain, John Chapple was the first to raise the alarm. In January 2007, he lost all of the 14 colonies in his back garden in west London. "It's too cold at that time of year to open the hives, so I always check on the bees by giving the hive a thump and waiting for what sounds like a roaring sound to come back," said Chapple. "But there was nothing, just silence." When he opened the hives to see what had happened to the bees, he found they had gone. Examination of a further 26 hives scattered across the capital revealed that two-thirds had perished.

"I was completely shocked," said Chapple, who chairs the London Beekeepers Association. "I could explain some losses to a failing queen or wax moth, but there were a few I could find no reason for. There was a queen and a few bees, but nothing else."

Chapple's enquiries as to whether the parks where he kept some of his hives had sprayed new pesticides drew a blank. Yet he was not alone. Other beekeepers in north-west London also reported strange losses. Chapple called the disappearance the "Marie Celeste syndrome". A year later, a survey of hives by government bee inspectors across Britain found that one in four colonies had perished.

There are around 250,000 honeybee hives in Britain, and a recent estimate by the department for environment, farming and rural affairs revealed that bees contribute almost £200 million ($279 million) a year to the economy through their pollination of fruit trees, field beans and other crops. In addition, the 5,000 tonnes (11 million pounds) of British honey sold in UK stores generates a further £12 million ($17 million).

Lord Rooker, the then farming minister, admitted in the House of Lords that honeybees faced a grim future: "If nothing is done about it, the honeybee population could be wiped out in 10 years," he said.

But despite the importance of bees to the nation's economy and the mounting evidence that some are dying in a mysterious fashion, the British government refuses to accept that CCD has reached its shores. As a result, it has turned down demands from the British Beekeepers Association for £8 million ($11 million) to fund research into bee health.

Similarly in Canada, although beekeepers across the country in 2007 reported average losses of 30% of their honeybee colonies – twice the normal winter rate – with 59% wiped out in the province of New Brunswick, the government has not attributed the cause of death to CCD. Instead, unpredictable cold spring weather, and inadequate controls for nosema and varroa have been blamed for the deaths.

In northern Europe, severe winter losses with almost identical symptoms to those being observed in the States have been reported since 2002. Borje Svensson has kept bees in Sweden for 30 years. In 2002-3, he lost 50 out of 70 colonies. "They were dead without visible reason," he recalls. "They had plenty of good sugar feed. Pollen store was available."

That winter, Sweden lost 38% of its bees, a fifth died in France, and 32% of Germany's million bee colonies were wiped out by a phenomenon known locally as *Kahlfliegen*.

Spain, which is home to a quarter of all Europe's nine million honeybee colonies, has also reported severe losses, as have Portugal, Italy, Poland, Austria, Belgium, the Netherlands, Slovenia and Croatia. Accurate figures, however, are hard to come by and no one knows how many deaths are actually associated with CCD.

In Germany, bee research institutes run a countrywide project to monitor 125 keepers' colonies. Losses were only 7% in 2004-5 and 13% the following winter, but a questionnaire sent to thousands more beekeepers revealed rates of loss that were twice as high. Other findings included nosema spores in more than 30% of colonies, bee viruses in a fifth, but traces of the neonicotinoid pesticide imidacloprid in only two out of 135 samples of soil, flowers, nectar and pollen. Peter Rosenkranz, chairman of the monitoring group, said the evidence after two years of work was too inconclusive to draw a clear correlation between "colony losses and certain factors".

The Swiss pathologist Dr Peter Neumann is not surprised by the Germans' lack of success. "Efforts by individual countries to reveal the drivers of colony losses are doomed due to the high number of interacting factors," he says. For this reason, Neumann, who works at Switzerland's bee research centre, is coordinating an international network examining honeybee losses. What started life as a European group now covers 35 countries, including the US, China and South Korea. He estimates that the cost of inexplicable and sudden bee losses to the European economy, even excluding pollination value, is more than 400 million euros (£375 million or $525 million) a year. But until the fledgling Coloss (Prevention of Honeybee Colony Losses) group had secured EU funding to replace patchy data across Europe with more reliable figures on bee deaths, the full extent of the problem would remain unknown.

Coloss – a kind of Interpol for bees – aims to share the latest information from scientists, apiarists and industry in order to track down the culprits behind the global bee deaths and prevent them from striking again.

French beekeepers who fought a successful battle to get two pesticides, Gaucho and Regent, banned in their country are still adamant that systemic chemicals used on crops are behind the spate of losses. Spanish scientists, meanwhile, accuse the Asian parasite *Nosema ceranae*, which wreaks terrible damage on the bees' internal organs. In 2007, scientists at the Regional Apiculture Centre in Marchamalo, near the central Spanish city of Guadalajara, discovered the new variety of nosema – a hardier version of its close cousin *Nosema apis* – in samples of bees from Germany, Spain, Switzerland and France. "We think that *Nosema ceranae* could do it [cause CCD] alone," said researcher Mariano Higes.

Professor Hans-Hinrich Kaatz of Germany's Halle University, meanwhile, points the finger at genetically modified, insect-resistant sweetcorn. He was director of a study between 2001 and 2004 examining the effects on bees of pollen from a genetically modified maize variant called Bt corn. It showed that while there was no evidence of a toxic effect of Bt corn on healthy honeybee populations, those infected with a parasite suffered a "significantly stronger decline".

A bacterial toxin in the GM corn may have "altered the surface of the bee's intestines, sufficiently weakening the bees to allow the parasites to gain entry," said Kaatz. "Perhaps it was the other way round. We don't know."

According to Neumann, there is "not one healthy honeybee colony" in the whole of Europe. The parasitic mite that ravaged bee colonies across Europe 30 years ago and has attacked them ever since has seen to that. *Varroa destructor*, as its grim name suggests, is an effective vector of viruses and diseases. Although Neumann doesn't believe in a "monocausal effect" for the current incidence of bee mortality, he is convinced that this mite plays a pivotal role. "It has to be the backbone of the problem," he says. His equation to solve the dilemma is: "Varroa, plus x, y and z".

In China, where all the honeybees in southern Sichuan province were killed by uncontrolled pesticide use in the 1980s, Professor Tan Ken, of the Eastern Bee Research Institute, says varroa is a major headache for the country's 200,000 beekeepers. Another mite, *Tropilaelaps clareae*, which has yet to reach Europe or the United States, also needs to be kept in check.

While he says it is not usual for China to lose as many as 20% of its seven million honeybee colonies during the winter, there have been no reports of a recent dramatic rise in bee deaths.

According to Professor Ken, many of China's bees are transported long distances, mainly for honey production. The country used to be by far the largest exporter of honey, sending more than 100,000 tonnes a year abroad until 2002, when the EU and US imposed a ban after the discovery of residues of chlordimeform, an antibiotic used to treat brood diseases. But it is still the biggest exporter of bee products such as royal jelly, which commercial beekeepers in the US use as a food supplement.

China is keen to cooperate with the Coloss group because the health of its massive honeybee population (three times that of the US) is vitally important for its fast-growing economy. When the bees were wiped out in southern Sichuan, farmers had to pollinate their pears by hand, a long, labour-intensive process that the rest of China could ill afford to follow.

It remains to be seen how successful an international bee working group can be when its members favour so many different theories. Neumann and many of his European colleagues are sceptical that CCD even exists. "My impression is that it is just a reinvention of the wheel of losses we experienced more than one hundred years ago," he says.

In his 1879 tome The ABC of Bee Culture, Amos Ives Root, a pioneer of US beekeeping, marvelled that "when a bee is crippled or diseased from any cause, he [sic] crawls away ... out of the hive, and rids the community of his presence as speedily as possible. If bees could reason, we would call this a lesson of heroic self-sacrifice for the good of the community." Neumann says that it used to be too cold for bees to leave the hive in winter to die, but now with milder weather they are able to fly or crawl away.

Denis Anderson, the Australian entomologist who identified *Varroa destructor*, is also unconvinced that CCD is a new disorder. In a letter to Science magazine in February 2008, he argued that "the symptoms are indistinguishable from those of the normal winter colony collapse reported in the United States since the late 1980s and attributed to nosema infection and/or the secondary effects of varroa." He also called on the authors of the Science report that linked CCD to the importation of Australian bees to issue a retraction.

When we meet Cox-Foster, however, she vehemently denies that CCD is nothing new. "What is terrifying about CCD is that the symptoms and the magnitude are totally different to anything we've seen in the literature," she says.

In an industrial greenhouse not far from the redbrick campus of Penn State University where she has her office, Cox-Foster, a warm, down-to-earth woman with white hair and a soft, youthful voice, is keen to show us the experiments designed to try to recreate CCD in bees.

They began in November 2007 when 20 2lb packages of varroa-free bees arrived from Hawaii. Confined to the greenhouse, they were fed a pollen substitute and a energy-boosting sugar drink. By the beginning of January they had built new comb and new brood and looked strong and healthy. At this point, 15 colonies were given a potion of sugar, water and Israeli acute paralysis virus (IAPV).

When we arrived six weeks later, we didn't need to wear protective bee suits to enter the greenhouse. There were just a handful of dead bees scattered across the floor of the four rooms. Rob Anderson, a young researcher working with Cox-Foster, sweeps up the bees every few days. He has marked a grid on the floor with masking tape, and notes how far from the hive the bees have died. A couple of the bees are lying on their backs twitching. When we flip one over, she cannot fly and tries to crawl along the floor instead, shaking. It is a distressing, but not uncommon, sight in sick bees. Bees infected with acute paralysis virus during the so-called Isle of Wight disease displayed similar shivering behaviour. And more recently when IAPV was discovered it was accompanied by shivering bees close to the hive.

We open a hive and see just a small cluster of bees around their queen on the top of a frame, much as the beekeepers who have witnessed CCD have described.

Although the IAPV experiment led to rapid die-offs and bees abandoning their hives, often showing shivering symptoms, Cox-Foster admits that it failed on a number of counts. One room of control bees somehow contracted IAPV and died, and most of the bees perished so quickly – within two weeks of getting the virus – that researchers had little time to observe how other factors such as poor diet or exposure to miticides (the chemicals placed in hives to kill varroa mites) affected the virus.

However, analysis by North Carolina State University of the proteins in dead CCD and non-CCD bees suggests that bad diet (or nutritional stress, as the scientists call it) was not a cause of CCD because healthy and ill bees contained similar levels of proteins.

Cox-Foster planned to start her greenhouse experiments afresh, with new IAPV-free bees. In the meantime, she and colleagues in Columbia University, the USDA and Israel have carried out more extensive sequencing of the virus and discovered three distinct strains of IAPV. One matches the original virus found in Israel, another is primarily in bees from the east coast and middle America, and the third strain is in west coast bees. "What these different lineages suggest is that there are different entry points into the US for the virus," says Cox-Foster. "The difference in their virulence and how pathogenic they are needs to be explored."

When we visited Cox-Foster, her team had just received $150,000 (£107,000) from Häagen-Dazs. Its total gift of $250,000 (£179,000) to fund bee research at two universities – the other was University of California, Davis – made headlines around the world. With 25 of its 60 flavours dependent on fruits and nuts pollinated by bees, the ice-cream maker wanted to stamp out CCD while at the same time raising consumer awareness about honeybees and its "caring" brand. A new flavour was to be launched called Vanilla Honey Bee.

Part of the money, which Cox-Foster had jumped through several hoops to secure, would be spent on two urgently needed pieces of equipment. One used liquid nitrogen and a ball bearing to smash bees to pieces much more quickly than grinding them up in a test tube by hand; the other allowed up to 800 bee samples to be analysed in a day. The remainder would pay for the university's findings to be disseminated on the web and for a programme aimed at teaching the public how to plant bee-friendly gardens and roofs.

Häagen-Dazs wasn't the only company to perceive bee rescue as an attractive marketing opportunity. Burt's Bees, the personal care products business, gave an undisclosed amount to create the Honeybee Health Improvement Project, a research task force. It also launched a public service announcement to run in cinemas showing Jerry Seinfeld's Bee Movie.

The animated film starring one of America's top comedians was released in the autumn of 2007 when the furore surrounding CCD was at its height. In the US, CBS's popular current affairs show, 60 Minutes, investigated the mystery, while PBS's Nature programme screened a documentary called Silence of the Bees. Around the world, news channels asked, "Where have all our bees gone?"

Pennsylvania state apiarist Dennis vanEngelsdorp, cofounder of the CCD working group, says the crisis caught the public's imagination because it makes sense that you can't have a world without bees. "In a world where there's

Ancient Egyptian
hieroglyphic featuring
the honeybee

Honeybee pollinating
a flower collects
pollen on its back
legs to feed its young

An almond grower attends to his beehives
among 600,000 acres of almond trees

(© Robert Yager)

At Penn State University Diana Cox-Foster
and Rob Anderson conduct experiments to
recreate colony collapse disorder

(© Alison Benjamin)

Multiple bees on almond blossom in California

Thousands of beehives being unloaded in February, in California's Central Valley, for the annual pollination of almonds

Returning forager bee does a waggle dance to
communicate the location of nectar and pollen

(© Scott Camazine/Alamy)

The blood-sucking, parasitic varroa mite
has killed billions of western honeybees

(© The Natural History/Alamy)

wars, global warming, and huge deficits that we don't understand, the basic person gets that without bees there's something wrong."

What struck us as odd when we met arguably the world's leading researcher on CCD was how the fate of the United States' multi-billion-dollar agricultural industry was in the hands of a few cash-strapped university research teams grateful for the odd corporate handout or federal grant.

The 2007 Farm Bill contained $76 million (£54 million) for honeybee research and conservation efforts to benefit native pollinators. But the bill had stalled on Capitol Hill. An additional request for $20 million (£14 million) by a bipartisan group of senators, led by California's Barbara Boxer and including Hillary Clinton, had fallen on deaf ears.

The US agriculture secretary, Ed Schafter, made it clear he didn't think more money was needed to study the problem. "You can always try to overwhelm things with money and hopefully get a better answer," he said. "Very, very seldom do you get a better answer; you're just wasting money."

What Schafter failed to grasp was that we don't yet have any explanation for CCD, prompting Kim Flottum, editor of leading US beekeeping journal Bee Culture, to ask: "If we don't have one [an answer] already, how can we get a better one?"

Schaffer thought the scientists – or rather one scientist – had CCD under control. "We have this great bee guy who's tinkering away in the lab to see what's going on," he said. In reality, the CCD working group is working hand to mouth, with funding from Florida and Pennsylvania's agricultural departments, the USDA, and $100,000 (£72,000) from the national honey board.

We flew to Penn State University, from the mechanised orchards of California, with visions of the same kind of industrial efficiency that went into growing almonds being applied to finding a cure for the stricken honeybees. We expected to see rows upon rows of scientists in white coats working round the clock in hundreds of huge laboratories. But nothing could be further from the truth. Cox-Foster's entomology department had just one small, antiquated-looking laboratory. In addition to the obvious lack of manpower and resources to tackle this national catastrophe, the timetable to which it worked failed to reflect the urgency of the situation.

CCD had now turned up in 35 states – according to a survey sponsored by Bee Alert Technology, a firm that sells hive-tracking devices – and wiped out a third of US bees, according to the USDA's Jeff Pettis, but the people who held the key to the killer's identity were having to wait months for papers of their findings to be published in scientific journals.

Given the scale of the problem, couldn't the investigation be speeded up?

"I'm frustrated by the speed we're doing, but advances take time," Pettis told a meeting of beekeepers.

So, a year after CCD was first detected, the crime remained unsolved, leaving beekeepers to go into the winter of 2007-8 unsure how many of their bees would pull through. As Hackenberg told Time magazine, and anyone else who cared to listen, "There's something ruining my livelihood that I have no control over."

To beekeepers pesticides are definitely part of the problem. "A lot of beekeepers blame neonicotinoid insecticides," said Troy Fore, executive director of the American Beekeeping Federation. "These are safer for humans and other mammals but they affect the neurological systems of bees. They don't kill the bees outright but they cause them to act in ways different to the norm."

According to research by entomologist Maryann Frazier and chemist Chris Mullen at Penn State, beekeepers are right to be concerned. They identified no fewer than 43 pesticides in 92 samples of pollen, 17 of them in one piece alone. To many beekeepers' surprise however, imidacloprid, the neonicotinoid banned in France for allegedly killing bees, only turned up in seven of the samples – and at low levels. The most frequent chemical residues were coumaphos and fluvalinate – used by beekeepers themselves in the hive to control the spread of varroa – and chlorpyrifos, a highly neurotoxic insecticide developed from a second world war nerve gas that is widely used on corn, cotton, citrus and alfalfa.

Coumaphos and fluvalinate also turned up at high concentrations in every sample of wax taken from a hive with CCD, and more frequently in bees with CCD than in healthy specimens. Most worrying, some levels of fluvalinate were reaching the lethal dose for bees.

"We're putting stuff in the hive to control mites which is highly toxic to bees," Frazier told a packed meeting of the national beekeeping conference in January 2008 in Sacramento. But she said that although a "weak trend" between CCD and in-hive miticides had been detected, it was too early to say whether this was the cause of CCD.

When we asked Mullen, during our visit to Penn State, why miticides would be a problem now, given that beekeepers have been treating their bees with these chemicals for 20-odd years, he replied: "There are lots more types of treatments taking place now because varroa has built up a resistance to traditional control methods. As a result there are more opportunities for abuses."

In addition, fungicides were found in pollen samples. These have been known to synergise with pesticides to become a thousandfold more toxic. Metabolites – breakdown products of pesticides that can be more toxic than the parent compounds – were also present. "This complex situation needs lots more work," Frazier concluded.

A similar conclusion was reached by all the researchers who presented their findings at the national beekeeping conference. "We can't point to one smoking gun," Pettis told the audience. "There are different parts of the jigsaw. We need to look at how it's coming together."

Such words provided little consolation to beekeepers whose worst nightmare had come true. After the horror movie that was CCD, they were about to experience CCD 2.

Of the 2,600 colonies that Hackenberg had at Thanksgiving, less than 1,000 made it through to 2008. Out of the 384 packages of Australian bees he bought to restock, only 25 survived.

"I shouldn't say I knew it was going to happen again, but knowing what I think I know I wasn't surprised," says Hackenberg. "You've got to look at what's new here. I've been trucking bees for 40 years, so that's not it. It might be one of the factors, but it's not the cause of CCD. We've had problems with mites and viruses for years. But something happened several years ago to start this problem."

In retrospect, Hackenberg thinks it began back in 2004. In May that year, blueberry farmers in Maine complained that his bees, which were pollinating their crops, were swarming and abandoning their hives. In addition, bees from other colonies were not stealing the honey that had been left behind.

Racking his brains for an explanation for this strange behaviour, he discovered that apple growers in Washington state had used a new neonicotinoid pesticide called Assail on their trees. His bees had been pollinating those apples in the spring.

That winter (2004-5) he lost almost a third of his bees – a much higher proportion than usual. The following year half died, and widespread losses were reported across the country.

"It got so bad, but no one could put a handle on it," says Hackenberg. So in the summer of 2006 he and a dozen other beekeepers held a meeting with scientists in Nebraska to try to find a reason for the rapid rise in bee deaths. "They say we had the best minds in the business at that meeting but we sat there for two days and knocked our heads against the wall and came away with nothing," he recalls. A few months later, two-thirds of his remaining bees vanished.

Beekeepers who trucked 1.2 million beehives to California's almond orchards in February provided the first indicator of bee health in 2008. The signs were not good. Of the dozen or so we talked to, only a couple had come through the winter relatively unscathed. The others had lost from 30% to 60% of their bees to what looked like CDD. Of the 12 migratory operations tracked by the USDA from September 2007 through to spring 2008, five reported abnormally high average losses of 44%. Of these, three had CCD-like symptoms; the other two were suffering from high varroa infestation.

One loss in particular sent shock waves through the industry. In the Californian desert, just north of Bakersfield, 40% of Adee Honey Farms' 70,000 beehives lay empty. The colonies that Richard Adee had told Congress a year earlier showed signs of dwindling had finally succumbed to the plague.

With CCD claiming the scalp of the largest beekeeper in the US, the stakes increased. The USDA announced a five-year programme, co-ordinated by Jeff Pettis, to "improve honeybee health, survivorship and pollination availability". In effect, the programme aims to pool expertise at the country's four federal bee laboratories.

Researchers at Beltsville, Maryland, are attempting to improve the longevity of honeybee queens, find effective controls for nosema and varroa and reduce the stress on bees of travelling long distances. In Weslaco, Texas, work is also focusing on varroa control and reducing migratory stress. At the honeybee breeding, genetics and physiology research lab in Baton Rouge, Louisiana, scientists are using genetic selection to create bees that are strong in the spring. In Tucson, Arizona, protein supplements are being developed that build up bees, and researchers are exploring how cross-breeding docile western honeybees with their hardy Africanised cousins could produce a disease-resistant superbee.

But isn't the five-year timescale too long? According to results from an Apiary Inspectors of America survey, the second year of CCD saw 36% of honeybee colonies die over the 2007-8 winter, a 14% increase over the previous year. That's up to a million honeybee colonies across 29 states.

Many beekeepers, like Hackenberg, could have gone out of business within five years. "It's like Kenny Rogers said: 'You gotta know when to hold and when to fold'," he says about his future.

Moreover, beekeepers warn that an ecological crisis caused by unprecedented honeybee losses is fast approaching.

Beeologics, an Israeli company with offices in the US, has responded to the emergency by developing an antiviral drug, which has yet to be licensed, that it hopes will protect bees from Israeli acute paralysis and other viruses. Its chief scientist, Ilan Sela, emeritus professor of virology and molecular biology at the Hebrew University in Jerusalem, is the scientist who discovered IAPV.

But is focusing on the virus the best way to tackle CCD? Isn't it better to address the causes rather than the symptom? If the bees' suppressed immune system is the crux of the problem, won't it be just a matter of time before another pathogen takes hold? Isn't the solution to be found in identifying what is making bees so weak?

That, however, could involve taking on the might of the pesticide companies and agribusiness.

CHAPTER 6

PESTICIDE POISONING AND GM CROPS

One of the main suspects in the current bee massacre was first sighted more than a decade ago in newly planted sunflower fields in central France. Here, beekeeping was enjoying a renaissance as hectare after hectare of countryside was swallowed up by the bright yellow flower whose nectar the bees rapaciously devoured.

French beekeepers had never had it so easy. But that was about to change. In July 1994, the honeybee population suddenly plummeted just as the sunflower nectar came on stream. Those foragers that had not "melted away", as the French put it, displayed very strange behaviour. They were seen taking long periods of rest on the sunflower heads and appeared agitated, constantly cleaning their antennae and scratching their bodies with their hind legs. Some fell to the ground and were hit by a form of paralysis, while those that made it back to the hive were often refused entry by the bouncer-like guard bees.

Mad bee disease, as it quickly became known, was soon reported in other regions of the country and accompanied by winter bee deaths of between 20% and 40%, compared with the 5-10% seen in previous years.

When beekeepers looked for an explanation for this catastrophe, they kept coming back to the same answer: farmers were using a new type of pesticide.

In 1994, imidacloprid was introduced as a seed dressing for sunflowers in the area of France where bee mortality later soared. An artificial form of nicotine, this pesticide acts as a neurotoxin, attacking an insect's nervous system on contact or ingestion. It is also systemic, meaning that it moves throughout the plant from the place it was applied, to the shoots, stem, leaves or flowers.

Manufactured by chemical giant Bayer under the trade mark Gaucho, this seed treatment is one of a new breed of pesticides classed as neonicotinoids, or "neonics" for short. Other products containing imidacloprid are sold all over the world for home and garden use, including Merit, for protecting turf, Premise for termite control and Advantage, a flea treatment for pets. Little did we realise until we started researching this book that we had been applying the chemical on the back of our cat's neck every three months without a second thought.

It turns out that imidacloprid is recognised as highly toxic to certain birds, including house sparrows, young fish and some freshwater crustaceans, such as mysid shrimps, that are a major source of food for many fish species. Moreover, it's acutely toxic to earthworms, one of the most important creatures in the soil ecosystem, and can readily leach through soil to contaminate ground water.

Like all pesticides, it fails to differentiate between good or bad insects. In lab tests, few spiny soldier bugs – a predator of potato beetle and corn earworm – survived a typical application of imidacloprid. The US Environment Protection Agency (EPA) categorises it as "highly toxic" for honeybees, which means it can kill upon contact as well as in residues. So why is it registered for use?

In its favour scientists point to its low toxicity to humans and other mammals. It is supposedly a farmer's best friend. In France, for example, Gaucho proved so successful at killing aphids that by 1997 farmers were using it to treat half the country's sunflower seeds.

Within a few years, however, the National Union of French Beekeepers (Unaf) reported that as many as a third of the country's 1.5 million registered bee colonies had disappeared, while honey production in south-west France fell by 60%. France's 75,000 apiarists angrily accused Gaucho of killing their bees. A long-running and bitter dispute began between apiarists, honey producers, environmentalists and

Bayer over the impact of the company's bestselling product. It was a high-profile row that saw beekeepers take to the streets in protest and would draw in the French agriculture minister and the high court, and have repercussions across Europe. Bayer even tried to sue one of the leaders of Unaf for disparaging its product, but the action was dismissed.

The company's own field and tunnel trials concluded that Gaucho "caused neither a reduced visitation of flowers nor an increased loss of foraging honeybees" and found "no records of behaviourally impaired honeybees". And Bayer fiercely contested the findings of a French agriculture ministry research programme that showed under laboratory conditions that a very small dosage of imidacloprid – a few parts per billion – could impair honeybees' learning performance. Bayer's data showed that the bees would never come into contact with even this low amount because "no quantifiable residues" of the pesticide were found in the parts of the sunflower visited by the bee when it was in blossom.

"All these studies show concomitantly that beehives exposed to Gaucho-treated sunflower fields did not develop any unusual symptoms," said Bayer scientist Rod Schmuck.

The French agriculture minister, however, was not convinced and, employing the precautionary principle, suspended the use of Gaucho on sunflower seeds in 1999 while further studies were carried out. Since finding after finding failed to rule out that the pesticide was completely innocuous for honeybees, the ban kept being extended. It is still in force today.

In contrast to Bayer's research, French studies found residues of imidacloprid in sunflower nectar and pollen at potentially hazardous levels for honeybees that "can affect their learning abilities" and impair their memory. And when individual bees were exposed to minuscule sublethal doses their foraging activity decreased and they became disorientated, which researchers concluded "can temporarily damage the entire colony".

A team of scientists led by the National Institute of Apiculture in Bologna discovered that at a high level of concentration – more than 100 parts per billion – the pesticide could cause a delay in bees' returning home, and at 500-1,000 parts per billion it was "likely that bees got lost and died somewhere in the field". This may be explained by the fact that some pesticides can cause a disturbance in the bees' dance language. Bees disappeared from the hive "probably due to the disorientation caused by the substance", the researchers concluded. However, it is highly improbable that bees would ever encounter such massive amounts in the field.

Bayer still maintains there is no evidence of any link between Gaucho and the drop in bee population, citing numerous independent field tests demonstrating that the sunflower seed dressing poses no risk during flowering.

Despite the ban, bee deaths did not noticeably decrease in France. Beekeepers blamed the continued use of Gaucho on sweetcorn seeds. Their fears were vindicated in a report, published by the French scientific and technical committee, that found that the risks to honeybees of corn seeds treated with Gaucho were "as alarming as with sunflowers". "The consumption of contaminated pollen can lead to an increased mortality of care-taking bees, which can explain the persisting bee deaths even after the treatment ban on sunflowers," it said.

In 2004, Gaucho was also banned as a corn seed treatment in France, a decision described at the time by Bayer as "incomprehensible".

The apian unrest spread to Bayer's home in neighbouring Germany, where beekeepers and environmental groups demanded an interim ban on imidacloprid products following the death of half the country's honeybees. But the German government was less receptive than its French counterpart.

It was not just Bayer that militant French beekeepers picked a fight with. Another German chemical giant, BASF Agro, manufactures another sunflower seed treatment called Regent that is also highly toxic to bees. It contains the neurotoxin fipronil and several studies showed that, like Gaucho, it contaminated pollen. After a fierce campaign, the government pulled the plug on Regent and a number of other fipronil-based products designed to control ants and protect lawns and golf courses.

The pesticides ban did appear to stem the massive bee die-offs. In 2006-7, winter bee deaths across France were less than 10%. "These results are intrinsically linked to the suspension of Regent and Gaucho," Unaf proclaimed. Bernard Vaissière, a researcher at Inra (France's national agricultural research institute) was also convinced of their culpability. "It is difficult to imagine that these insecticides had no impact. They were in the pollen and the nectar. I don't see how the pollinating bees could have failed to ingest them."

Yet honeybee mortality rates across France shot back up to more than 60% for 2007-8, leading people to question again the role of pesticides.

Undeterred, French beekeepers are now targeting Cruiser, a neonic produced by the Swiss agrochemical group Syngenta. It has been approved for use as a treatment on sweetcorn seed intended for animal feed. But its active ingredient, thiamethoxam, is highly toxic to honeybees and is accused by Italian apiarists of killing their bees. Inra studies have shown how the flight behaviour of bees, especially their return to the hive, could be adversely affected by the absorption of very low doses of thiamethoxam.

For a pesticide to be authorised for use in Europe or America, the manufacturer is required to conduct LD50 tests – a standardised measure for expressing and comparing the toxicity of chemicals – to determine the dose that kills half the animals tested. You don't need much of the neonicotinoids or fipronil to achieve the LD (lethal dose) 50 for adult honeybees. But although the US Environmental Protection Agency (EPA) acknowledges this fact, it says it has never been a reason to ban the pesticides. "Some pesticides such as insecticides are intended to target insects and may be toxic to bees. If bees are likely to be exposed to a pesticide, then bee toxicity tests are required. Depending on the outcome of those tests, EPA classifies the pesticide as non-toxic, toxic, very toxic or highly toxic to bees and requires specific label language to restrict the use of the pesticide and help prevent exposure of the pesticide to the bees," says an EPA spokesman.

Critics argue that the LD50 is an insufficient test for this new breed of pesticides, whose sublethal effects on adult honeybees, brood and the colony as a whole need to be measured. They point out that the highly social honeybee acts not as an individual creature but as part of a well-ordered household. In a letter to Markos Kyprianou, the European commissioner of health and consumer protection, in November 2006, beekeeper associations and environmental organisations from across Europe called for imidacloprid, fipronil, thiamethoxam and chlothianidin – another neonicotinoid found in a Bayer sweetcorn seed treatment called Poncho – to be banned because they considered "the risk assessment of these active substances insufficient".

The letter drew attention to their "acute toxicity" for bees. For this reason, it asked that "no molecule with a high toxicity towards bees, in particular fipronil and imidacloprid, is registered as long as independent and validated tests have not shown the innocuousness of the product for bees, their brood, and the functioning of the colony."

The demand was reiterated by German Green MEP Hiltrud Breyer, who tabled an emergency motion in June 2007 calling for systemic neonicotinoid pesticides to be banned in Europe while their role in killing honeybees was thoroughly investigated.

Although both attempts failed to stem the tide of bee-toxic pesticides sweeping across Europe, proposed legislation could do just that. In 2008, the European Parliament agreed on its first reading of the regulation on the authorisation of pesticides that in future pesticides will not be approved for use in the EU if they are toxic to bees. Evidence will be based on "available data and information, including a review of the scientific literature".

The European Crop Protection Association, a lobby group for the agrochemical companies, warned that the proposed legislation could affect up to a fifth of the 210 most important substances on the market. It was, however, subject to approval by the Council of Ministers and, according to insiders, unlikely to get the go-ahead.

Across the Atlantic in Canada, beekeepers from New Brunswick, Nova Scotia and Prince Edward Island raised the alarm about imidacloprid not long after the French. In 1999, they lost up to half their honeybees after they had been foraging on land treated with Admire, a potato-protection treatment manufactured by Bayer. Although bees don't pollinate potatoes, they feed on clover and oilseed rape, which are grown in rotation with them. Ontario and Manitoba bees were similarly hit in potato areas using the pesticide.

Stan Sandler, the largest commercial bee operator in the Canadian Maritimes with 3,000 hives, witnessed a rise from an annual 5-15% bee loss to 60%. In 2001, he presented his concerns about low levels of imidacloprid causing what he described as "a sort of bee Alzheimer's" to Prince Edward Island's pesticide advisory committee. "The hives had what are now recognised as CCD symptoms," he recalls. "The foragers didn't come home and brood starved because there were no bees to feed them."

In response to the controversy, the Prince Edward Island and New Brunswick governments, together with Bayer, funded research in the Canadian Maritimes. An initial study by Jim Kemp, a botanist at the University of Prince Edward Island, involved gathering soil and clover samples from potato fields around the province, then collecting bees from nearby hives and extracting from them pollen and nectar that they had eaten. Traces of Admire were found in the soil and clover leaves, but not in the clover flowers, the pollen or the nectar. Kemp said this showed that although Admire residues may remain in potato fields, and even in clover plants, for years after the chemical is applied, there is no proof that it is being ingested by bees. "It means it's very unlikely that Admire is the cause of the problem," he said.

The Canadian Honey Council, sceptical that Admire could be ruled out as a bee-killer following one summer of research, called for the government to delay the registration of the pesticide until "unbiased" research had been conducted on its long-term, sublethal effects on bee colonies.

Further research part-funded by Bayer supported Kemp's original findings. A comprehensive mapping exercise over 800 square kilometres (310 square miles) where bees from 20 apiaries were monitored feeding on blueberry and clover fields during two summers found so many factors affecting the bees' health, says Kemp, that it was impossible to pin the blame on imidacloprid.

"There were pesticides, fungicides, foulbrood disease and mites," he told us. "We couldn't see imidacloprid in the pollen or the flowers."

Sandler, whose bees were among those studied, finds this hard to believe. He says he has data a year after Admire was injected into the soil, showing residues of three parts per billion found in oilseed rape pollen and honey.

"You'd be a fool to say that pesticides can't be a problem," Kemp comments. "But I didn't see a connection between imidacloprid and bee deaths. Bees were going down whether they came into contact with it or not."

A keen ecologist who is trying to increase native pollinators, such as bumblebees, on Prince Edward Island, Kemp believes honeybees are to the environment what canaries were to the coal mine. "They are trying to tell us something," he says, "but it's not: 'Imidacloprid is killing us.' "

Since the studies, farmers have started applying Admire as a seed treatment rather than injecting it into the soil or spraying it, so bees are less likely to come into contact with the substance. As a result, Sandler is now less concerned about Admire's effect on bees than he is about the lack of independent research being conducted in Canada. With universities expected to get industry funding to unlock government funds, he fears that truth is being sacrificed to profit.

Kemp agrees it could be a problem. "You have to be very cautious," he warns. "There are lots of cases where researchers and industry have worked too much on the same side." But he insists Bayer was hands-off. When he asked Bayer how it would respond if he found anything it didn't like, he says, the company told him: "We have to know, because we've got to solve it."

Peter Kevan, a biology professor at the University of Guelph, claims, however, that Bayer has turned down his proposals to investigate sublethal doses of imidacloprid on bees' behaviour. "I can't do research that asks questions that need to be answered," he says.

Similar obstacles have been reported in Germany, where Hans-Hinrich Kaatz, a professor at Halle University, tried to get funding to continue his studies on the potential impact of genetically modified sweetcorn on honeybees. Although honeybees don't pollinate sweetcorn, they do collect its pollen for food. His initial project, which we described earlier, showed that while there was no evidence of a toxic effect from a genetically modified sweetcorn variant called "Bt corn" on healthy bees, there might be an effect on bees infested with the gut parasite nosema.

According to the study there was a significantly stronger decline in the number of infested bees that had been fed a highly concentrated Bt feed. Although the feed was administered in the lab over six weeks – longer than the bees would normally be exposed to it in the field – and at higher concentrations than foragers would find, Kaatz thought his rather skewed findings warranted more study. But, he told the German magazine Der Spiegel: "Those who have the money are not interested in this sort of research, and those who are interested don't have the money."

*

But why such concern about genetically modified crops in the first place?

GM crops are now grown in 40% of sweetcorn fields in the United States and two-thirds of China's cotton plantations. All of these crops have had a small amount of genetic material from other organisms added to them in order to protect them from pests, reduce their vulnerability to herbicides, or change their characteristics in some other way.

One particular type is Bt crops where a gene from the bacterium *Bacillus thuringiensis*, which has natural insecticidal properties, is inserted into the corn. Bt crops, which include potatoes, sweetcorn and cotton, produce insecticides in every cell, reducing the need for traditional chemicals.

Understandably, perhaps, there are concerns about how this might affect honeybees. The US National Research Council's influential report on the status of pollinators in North America, published in 2006, quotes a review of the literature on GM crops that concluded that "in some cases there are sublethal effects attributable to consumption of transgenic pollens".

Joe Cummins, emeritus professor of genetics at the University of Ontario and a member of the anti-GM Institute of Science in Society, goes further, arguing that Bt biopesticides, and neonicotinoids that are frequently used as dressing on GM seeds, enhance the killing power of parasitic fungi that can be found in honeybees. He has presented evidence that sublethal levels of imidacloprid can act synergistically with parasitic fungi such as nosema to increase 45-fold the toxicity of the protein released by GM plants.

"There is a good indication that exposure to sublethal levels of systemic insecticides used in seed treatment of both conventional and GM crops and in widespread soil and foliar applications can affect beneficial insects by reducing their immunity to parasitic fungi," says Cummins.

Since *Nosema ceranae* has been found in the guts of CCD-affected bees (although it is also present in healthy bees), might Cummins be on to something? His findings seem to chime with the work started by Professor Kaatz in Germany.

The scientific consensus, however, is that Cummins, an outspoken critic of GM foods, is overstating his case, and there is scant evidence to link GM crops with disappearing bees.

Jeff Pettis, research leader of the United States Department of Agriculture (USDA) honeybee labs and joint chair of the CCD steering group, took part in recent field studies that showed no evidence of any of the lethal or sublethal effects of Bt proteins on honeybees that Cummins refers to. "We directly fed the Bt toxin to whole bee colonies and could demonstrate no effects on them," Pettis told a round table of bee experts.

Moreover, the link between GM crops and CCD appears particularly flimsy when comparing states that grow a lot of GM corn, such as Kansas and Nebraska, with states that reported cases of CCD in 2007. They don't match.

The Green MEP Hiltrud Breyer, however, extensively cited Cummins's findings in her emergency motion to the European Commission on protecting the honeybee. The way to halt colony collapse, she stated, was to ban Bt crops as well as neonics "while their synergistic action in killing honeybees in combination with parasitic fungi and other infections are thoroughly investigated".

Of course, bee deaths from pesticide poisoning happened long before the invention of genetically modified crops or the latest breed of pesticides. We were shocked to read just how casually and clumsily millions of honeybees have been needlessly killed by pesticide sprays and applications ever since they were first used. In her seminal book Silent Spring, which warned the world about the effects of agricultural chemicals on the environment, Rachel Carson described how beekeeping as an industry nearly died out in the arsenic-sprayed cotton country of the southern US at the beginning of the 20th century.

After the second world war, when synthetic pesticides were produced as a by-product of chemical warfare, bee kills became even more commonplace. There were two types of these man-made elixirs of death: chlorinated hydrocarbons, of which DDT was the most widely-used, and organophosphate insecticides, such as parathion. Contact with parathion, which like today's neonicotinoids targets insects' nervous system, made honeybees become "wildly agitated and bellicose", perform frantic cleaning movements and become near death within half an hour, wrote Carson. Yet by the early 1960s some seven million pounds (3,000 tonnes) of the chemical was being sprayed on fields and orchards throughout the States, and professional beekeepers in California were reporting estimated annual losses of a fifth of their bees to pesticide poisoning.

Despite the creation of the EPA in the 1970s and its labelling requirements for pesticides, accidents, careless application or failure to adhere to the labels' recommendations and warnings meant that pesticide-attributable bee deaths continued.

Bob Harvey, an unassuming middle-aged man with a weather-beaten face, is an unlikely hero. But in 1997, this New Jersey beekeeper won a lawsuit after a quarter of his 3,000 bee colonies were killed by the pesticide called Penncap-M, which was being too liberally applied. It was particularly insidious because the active ingredient, methyl parathion, was enclosed in a tiny capsule that bees mistook for pollen grains and carried back to the hive to store. When eaten, it killed both the adult bees and the brood.

"In one county none of my bees survived," Harvey recalls. "We figured the farmers were using this stuff which had just been newly marketed. We managed to get all but one of them to stop using it and we took legal action against the farmer who wouldn't quit. When the farmer heard we'd got hold of his spray records and he was making six or seven applications, instead of the recommended four, he settled out of court for more than $600,000."

Unlike the old-style organophosphates, neonics don't kill bees the day they are applied. For this reason, beekeepers in the US couldn't finger them straight away. But Chris Charles and Dale Bowers are among nine beekeepers from North Dakota who are so convinced that Gaucho killed $40 million (£29 million) worth of their bees back in 1995 that they are taking Bayer to court armed with lab tests that, according to Charles, identify high concentrates of imidacloprid in the pollen of oilseed rape whose seeds were treated with the pesticide. "It also showed up in the wax, honey and dead bees collected over a three-year period," he says.

In a warehouse in Carrington, North Dakota, Charles has 30,000 disused hives as a reminder of the thousands of colonies that died. "I lost all 8,000 that year and plenty the next," says Charles.

A beekeeper since the age of nine, Charles says he's out with his bees every day, seven days a week, so he knows if they are acting strangely. "At first, I'd call them drunk bees: they didn't know what to do. Then thousands died," he recalls.

"If you go to the doctor to get your blood checked, you're going to believe the tests. So you've got to believe the tests that we did."

Seven years after they began legal proceedings against Bayer, the beekeepers are still awaiting the ruling that was expected in January 2007. Charles suspects the delay is because of the massive profits at stake if neonics are found to be at the heart of CCD. "It's frustrating that scientists and companies know what is causing it [CCD] but over money refuse to do anything about it," he claims.

Jerry Hayes, the Florida state apiarist who is a vocal critic of current pesticide use, says: "These things do a great job on termites: they can't remember how to get home, they stop eating and then fungus takes over and kills them … Termites are social insects just like the honeybees. So if you extrapolate to honeybees, this is kind of a scary thing. It amazes me the disconnect that chemical companies have – or are allowed to have – in terms of the effect [of pesticides] on good insects."

Dave Hackenberg, the beekeeper who discovered CCD, is less surprised: "Big Ag has control of the USDA from the Secretary right on down to the lowest guys on the totem pole," he says.

This attitude is rife among beekeepers we met in the

States, who position themselves as the small guy against the system. Hackenberg compares himself to Erin Brockovich, the legal assistant who brought down a power company accused of polluting a city's water supply. We joke about who would play him in the Hollywood film. Our money's on Tom Hanks.

Although the first report of CCD came out of Hackenberg's Florida yard in November 2006, he says he noticed something was wrong two years earlier after his bees had been pollinating apple trees treated with the pesticide Assail, which contains the neonic acetamiprid. In March 2007, he wrote to all his customers requesting them not to use neonicotinoids on the crops his bees pollinate.

"I am asking you as a grower to take a look at what you have used last year and what you might be using this year. If at all possible, please try to use something beside these products," he wrote.

"Two of my largest apple growers called me when they'd read the letter and said, 'We're part of your problem because we sprayed Assail right in the middle of the bloom,'" Hackenberg recalls. "The label says it can be used during bloom, but in small print and parenthesis – that shows how important they think it is – it adds 'except when bees present'."

His largest customer, the Maine-based Jasper Wyman and Sons Blueberry Co, was the first to agree to his request.

Other growers have followed, but few scientists are willing to put their money on pesticides. As with GM crops, they cite evidence of states, such as Arizona, covered in pesticide-treated vegetable crops, that had no reported cases of CCD in 2007.

Scientists, however, don't have a great track record when it comes to pesticides. The list of toxic chemicals that have been withdrawn from sale after the risks to human health and the environment became too obvious to ignore is endless. In 1972 DDT was banned in the US because of "human cancer risks" and "environmental hazards", according to the EPA. Other chlorinated hydrocarbons such as polychlorinated biphenyls (PCB) soon followed. They have been "demonstrated to cause cancer as well as a variety of other adverse health affects on the immune system, reproduction system, nervous system and endocrine system", the EPA notes. More recently, the residential use of home and garden products containing chlorpyrifos has been phased out due to its health risks to children.

Announcing the removal in 2000, of 800 products containing chlorpyrifos, which was widely used on the lawns of schools and residences, and in homes for termite control, the EPA said: "It belongs to a family of older, riskier pesticides called organophosphates, some of which date back 50 years or more. The time has come to ... where the science dictates, remove those chemicals that pose an unreasonable threat to human health and move to newer, safer alternatives."

According to Dr Carl Johansen, a retired Washington State University entomologist, scientists thought the first formulations of Penncap-M looked pretty benign. "In fact, we thought encapsulation would be less dangerous, but we found out differently later," he said. "Penncap-M is the worst thing we've had to deal with in terms of bees."

What if a farmer mixes together different neonics, or a pesticide and fungicide? "No one has ever tested what happens to the toxicity of these products if they do, simply because you're not required to," says Hackenberg. "Bayer's not going to say you can mix it, but people do."

Dave Ellingson, a Minnesota-based beekeeper whose bees went awol after exposure to CruiserMaxx – a combined insecticide/fungicide treatment – on soya bean flowers, told the US national beekeeping conference in January 2008 that farmers are "stacking" – combining insecticides, herbicides and fungicides in a single trip across the field. This combination could be 1,000 times more lethal than if the chemicals were applied separately, according to one study by the University of North Carolina, which looked at how certain fungicides synergised with neonicotinoids.

Researchers at Penn State University are currently investigating the potential link between neonics and CCD as part of the USDA-led CCD action plan. At this year's national beekeeping conference it was standing room only when entomologist Maryann Frazier presented her team's findings. Although imidacloprid was found in only seven of the 92 samples of pollen, and then at a low level of concentration, Frazier told her audience that the levels were still higher than had ever been reported before in the field. "It was found at higher levels than representatives from Bayer told us it would be. They were surprised," she said.

More controversially for beekeepers, the researchers uncovered a toxic soup of chemicals used in the hive to control varroa mites.

The miticides, fluvalinate and coumaphos, were detected at high levels in more than half of pollen samples and in every single wax sample. Polymer strips infused with slow-release fluvalinate, specially designed for use in beehives (sold under the name Apistan) have been used as a major varroa control treatment since the parasite was discovered in the States 20 years ago. But the mites built up a resistance. So coumaphos, an organophosphate spawned by the creation of nerve gas agents in the second world war, was added to the hives.

Frazier and her team were shocked to find this old-fashioned chemical, a common cattle dip, whose use in hives in the US they thought had stopped years ago. But with beekeepers reusing combs year after year, the residues build up.

One problem, says Frazier, is that pretty harmless chemicals have gradually become more toxic. Fluvalinate, for example, was registered by the EPA with a LD50 of 65.85 microgrammes. Today the formulation has changed and just 0.2 microgrammes will kill half the bees. Moreover some formulations of the product add stabilisers that can increase toxicity to 0.00964 microgrammes.

Some of the samples Frazier and her team analysed were sent by Penn State apiarist Dennis van Englesdorp, whose CCD investigations had taken him up and down the east coast to monitor colonies of bees. In Maine, he found something in hives he had never seen before: pollen "entombed" in a thick layer of black wax, as if to keep unwary bees away. More than a third of the colonies in Maine had this unusual black-capped pollen. By November 2007, 44% of them were dead, compared with just 20% of colonies without it. Tests revealed high levels of fluvalinate and coumaphos in the off-limits pollen grains.

Although the tests identified a link between high concentrations of miticides in wax and pollen and CCD, Frazier stresses that they can't prove it's the killer. "No way can we say this is the cause of CCD," she says.

One reason is that, oddly enough, a few of the healthiest colonies they tested had the highest levels of coumaphos not only in the pollen and wax, but also in the bees themselves. Frazier says they need to analyse many more bees before drawing any conclusions. VanEngelsdorp has samples from the 260 migratory colonies he has been tracking over a year waiting in freezers to be analysed, but there is no funding.

Work undertaken by Frazier's colleague at Penn State, entomology professor Diana Cox-Foster, however, may take researchers a step closer. She has found a link between a miticide and a honeybee virus. Using coumaphos to control varroa mites amplified a virus in the bees called black queen cell virus (BQCV), and killed them.

This research treated three samples of bees with different types of miticides: coumaphos; formic acid pads left in the hive for a period of time; and a higher concentration of formic acid administered in a "flash method" for 24 hours. The last treatment, though not yet registered as a varroa control with the EPA, is thought to be used by some beekeepers, Cox-Foster told us. A fourth sample was not treated for varroa and acted as a control.

While coumaphos and formic acid on pads were the most effective ways of keeping mite numbers extremely low, BQCV increased, and it was these colonies that died.

BQCV, which kills queens before they are born and blackens the cell in which they are growing, was discovered in South Africa in the late 1990s. Little is known about it, except that it is closely related to the IAPV virus, whose identification in most CCD-affected colonies in 2007 has been the only breakthrough in the CCD investigations to date.

"Normally it [BQCV] probably doesn't cause too much problem but the findings here show that these pesticides have an effect on the virus – making it more prevalent," Cox-Foster explains. "Somehow the treatment stressed the bees and caused the virus to amplify."

Back in France, keenly awaited results of a long-term study on bee mortality by the French food safety agency Afssa were finally unveiled in February 2008, and seemed to put paid to the long-held notion that imidacloprid and fipronil cause bee deaths. The study was conducted from 2002 to 2005 in 120 bee colonies from 24 apiaries throughout France. It looked at the interrelationship of factors such as pesticide use, viruses and beekeeping practices on bee mortality and revealed no significant differences in death rates between the years when Gaucho and Regent were used and after it was banned.

A close look at the study shows that imidacloprid and its metabolite, 6-chloronicotinic acid, were the most frequent chemicals found in the pollen, honey and bees, while fipronil and acetamiprid were also found at very low concentrations. But researchers say there was no evidence of any statistical relation between the presence of residues and colony deaths.

Like the Penn State study, they also found frequent coumaphos residues. Although the official veterinary product containing the miticide, Perizin, was not available in

France at the time of the study for controlling varroa, the report noted that home-made preparations using a flea treatment for dogs were applied in the hives instead, often at much higher doses than those recommended. "Some of the unexplained colony mortality may find its explanation here," it concluded.

Bayer has always maintained that bee problems in France and other countries have many causes. Following the study, it claims opinion in France has now shifted its way. "Today, no serious French scientist, politician and not even Michel Barnier, the minister of agriculture, claims that Gaucho is responsible for the bee mortality in France," said a company spokesman. Barnier has told the National Assembly: "Bee deaths have been scientifically confirmed in large-scale farming areas where products often claimed to be responsible, such as Regent or Gaucho, had not been used."

Bayer and Regent's manufacturer, BASF Agro, expected the ban on their products in France to be lifted soon.

Meanwhile, the hunt for the CCD killer continues.

Chlorpyrifos, the most widely used household pesticide in the US before its virtual elimination for home and garden use in 2000 on grounds of public health, was found in 61% of the pollen samples analysed by Penn State University. First registered in 1965 by Agent Orange creator Dow, it is classified as acutely toxic to honeybees but is still used on sweetcorn, cotton, citrus fruits and alfalfa, and applied to turf on golf courses. From 1987 to 1998, the EPA estimated that between 21 and 24 million pounds (roughly 10,000 tonnes) of chlorpyrifos was applied to more than eight million acres (3.3 million hectares) of crops in the US. Its use has been eliminated on tomatoes and restricted on apples and grapes because of risks to children. Could an old-style pesticide, rather than the neonics, be the killer?

The EPA says applications of chlorpyrifos would be expected to pose a risk to bees present in the treated area during application. "Results from some field studies confirm predicted risks to bees, which are killed if presented during application and for as long as 24 hours after treatment," it states. However, it has no information on what residues in pollen could do to bees.

Pettis is concerned about bees' low-level exposure to both chlorpyrifos and two fungicides also identified in pollen samples tested by Frazier and her colleague, Chris Mullen.

Myclobutanil, while found in only 12% of the samples, was at high levels where present. Likewise, chlorothalonil was only in 9% of the pollen but again at high levels. Both widely used to control fungi that threaten vegetables, fruits and turf, they are sold in garden centres under brand names such as Eagle, Nova, Bravo and Echo. Chlorothalonil is also used as a preservative in paint. The EPA, however, does not require fungicides to be tested for toxicity to bees.

"They could synergise with other insecticides to become

much more potent," says Pettis. "We just don't know." He hopes to carry out some further experiments this summer.

VanEngelsdorp says that pesticides would be a convenient villain for everyone – although the chemical companies may take issue with that. But so far such a link is not supported by the data.

He also warns against blaming beekeepers, who in their fight to control the varroa mite with a variety of chemicals could have unwittingly triggered these latest deaths. It's a classic Catch 22: don't use chemical and risk seeing your bees dying of varroa infestation (which has claimed millions more colonies across the world than CCD); do use them and risk the chemical build-up damaging the bees in the long term. As Jerry Hayes in Florida says: "We're trying to kill a little bug on a big bug."

Attention is now shifting to the role of the little bug that has brought death and destruction to *Apis mellifera* for decades. As *Varroa destructor*'s resistance builds to whatever is thrown at it, could it have spawned an even deadlier version that has made the bees disappear?

CHAPTER 7

THE ENEMY WITHIN

A great feat of early 20th-century engineering that connected western Russia with Mongolia and China could be responsible for the disappearance of honeybees on another continent more than 100 years later. The trans-Siberian railway allowed Russian beekeepers to transport their western honeybees (*Apis mellifera*) to countries where a sibling species, the eastern honeybee (*Apis cerana*) resided. The bees appeared to live side by side harmoniously, but the indigenous variety was host to a parasitic mite that has brought death and destruction to millions of western honeybees.

Varroa is a species of blood-sucking mite closely related to the tick. It happily co-existed with its original host, which over millions of years had developed ways of controlling the mite population so that it rarely caused it harm. But the western honeybee had no time to build up defences against the parasite. When the bee was exported back to Russia, it was infested. By 1953 the first case of varroa infestation was reported in Soviet Russia. During the 1960s the mite quickly spread to Hong Kong, the Philippines, China, India and Japan. The following decade it invaded eastern Europe, from Czechoslovakia to Bulgaria, Poland and Romania. There were also sightings in the South American countries of Paraguay, Brazil, Argentina and Uruguay. In all of these cases the mites had hitched a lift on the back of the hapless bees as they were moved around the world by man.

The canny mites that adapted to live on the western honeybees were discovered just a few years ago to be a distinct species from the *Varroa jacobsoni* that had colonised their cousins. To complicate things further, two strains of this newcomer *Varroa destructor* were identified: a dangerous Korean strain that inhabited western honeybees in Europe and the US, and a less harmful Japanese/Thai variety that lived on them in South America, where fewer varroa-infested bees have perished.

Under a microscope, *Varroa destructor* looks like a cross between a jellyfish and a Frisbee, with hairy legs. To the naked eye, the reddish-brown dots, which are about 1.1mm (¹⁄₂₄in) long and 1.5mm (¹⁄₁₆in) wide, are hard to make out on adult bees but clearly visible against the pale larvae on which they also feed.

To get an idea of what it must be like for a honeybee, which after all is only 12mm (½in) long, to have one of these mites clinging to her, we were told to imagine carrying a monkey on our back.

In the autumn of 1977, Wolfgang Ritter was a young PhD student at the Oberursel bee institute at the University of Frankfurt, where varroa was first spotted on *Apis mellifera* in western Europe. Russians had informed the institute's director that the eight-legged mite had had a dramatic effect in their country, but Ritter tells us they had no idea of the scale involved. "We told all the bee institutes in Germany what we had found and how it might be a problem, but we didn't know how or why," he recalls.

At that time little was known about the bees' adversary other than that it sucked their blood and killed many of their colonies. The institute's director asked Ritter to stay and research the parasite. "He said it might be one or two years' work," Ritter laughs. Thirty years later he's still busy.

"I'll tell you another funny story," he says. "When a group of Austrian researchers visited the institute early on, the director told them to make sure they saw my work because they might never have another chance to see varroa."

A few years later the mite decimated Austrian bee colonies as it continued its march across Europe. Today, Australia is the only varroa-free continent.

In Germany, where varroa rapidly spread, beekeepers

had three options: to kill all infested colonies, control varroa with chemicals, or leave the bees untreated and see how many survived. They opted to control and study the bees' foe. As a result, we now know about the mite's life cycle and peculiar sexual proclivities.

The adult female varroa mite lives either on an adult honeybee or within a wax-capped cell, where the larva develops into a bee. She enters the cell to reproduce just before it is capped and hides in the larval food. About four hours later, when the larva starts to spin a cocoon and eat the food, the mite feeds on the pupae's haemolymph (a blood-like fluid) and establishes a feeding site on her that her offspring can eat from as they develop. About 60 to 70 hours after capping she lays her first egg, which develops into a male. Five more female eggs are laid at 30-hour intervals. The male emerges five to six days later and the females a couple of days afterwards, to begin an incestuous affair on a mound of their mother's faeces, as the brother mates with his sisters, starting with the eldest.

Mites prefer to breed in cells containing drone larvae because drone bees take longer to develop. This gives the young female mites more time to mature and mate. In the more numerous worker bee cells, the younger sisters fail to reach puberty by the time the honeybee is ready to emerge and they die in the cell with their brother. The mated female mites, meanwhile, emerge with the young bee and hop on to other bees in the hive to feed. Using specially adapted mouth parts, they pierce their new host's cuticle and suck its blood. They then begin the breeding cycle all over again.

During the summer, varroa mites live for up to three months, during which time they can breed three or four times. With heavy infestations, two or more female mites may enter the same cell to breed. In winter the mites survive for longer on the bodies of the adult bees, which cluster in the hive until brood rearing commences the following spring.

Initially, scientists across Europe thought it was the mite's vampire-style feeding habits that damaged and killed the bees. But they couldn't understand why the damage was not proportional to the number of mites a colony suffered. In the early days of the varroa plague, it was not unusual for colonies of 50,000 bees to be infested with 15,000 to 20,000 mites yet survive. Others, however, with much smaller number of mites might die.

It was a British bee virologist, Brenda Ball, who solved the mystery. In 1986 she visited the Frankfurt bee institute to analyse dead bees for pathogens. As we have already discussed, she examined dead bees that were lightly, moderately and heavily infested with varroa, and found that most of the dead bees in heavily infested colonies had the disease called acute paralysis virus (APV).

Bees, like humans, carry a number of latent viruses. What made the virus lethal was the mite's feeding activity. If there was an active infection in one bee that got into its bloodstream, then when a mite fed on it and then its sisters it transmitted the virus from one bee to another, much as HIV/Aids can be spread by using dirty needles. In a few days the bees would be dead. Dr Stephen Martin, a varroa expert at Sheffield University, says the reason the parasitic mite killed billions of western honeybees, and became the most deadly honeybee pest ever seen, was that it created a "new viral transmission pathway". "As it spread, it altered the naturally occurring transmission pathways of honeybee viruses," says Martin.

The mites' original host, the eastern honeybee, had developed strategies to limit the mite population to just a few hundred a colony. It allowed the mites to reproduce only in drone cells, for example, by quickly removing its eggs from any other cells. Because the western cousin had not had time to perfect such defence mechanisms, Martin explains, the mite population increased 1,200- to 2,000-fold each year. This leads to thousands of what he describes as "viral vectors" living in a colony.

With 14 identified honeybee viruses, from APV (a five-day killer) to the closely related slow paralysis virus (SPV), which takes 12 days to kill bees, and the less virulent deformed wing virus (DWV), the mites have quite an arsenal at their disposal.

Stumpy bodies and shrivelled wings are the outward

signs of DWV, but it also slightly shortens adult bees' life-span. This is important to know, because if adult bees go into winter with the virus they can die before enough new bees are born in the spring to sustain the colony. "As a result, a small but significant proportion of over-wintering bees die prematurely and the rest of the colony gradually disappears as the balance of the colony is shattered," says Martin.

He believes that this imbalance has caused the symptoms reported as CCD in the States. The reports of hives that are empty save for a queen and a few workers, he says, match exactly what you find when old bees have died prematurely because of varroa before new bees have been born.

Martin was talking about honeybee colonies collapsing long before the term was coined in the States. For this reason, he disputes that CCD is anything new. "I'd bet my bottom dollar that varroa has got a lot to do with it," he says.

Americans are all too familiar with the devastating impact of the honeybees' parasitic mite. Its arrival in the country in 1987 sent shockwaves through the beekeeping community. "The introduction of the Asiatic bee mite is a nightmare come true for the North American beekeeping industry. Even as I write this, many persons are in a state of shock," wrote Malcolm Sanford, then an extension (public health) entomologist at the University of Florida. "There is near unanimous support that it is potentially the most serious pest ever to threaten US beekeeping."

Florida's commissioner of agriculture placed a two-week moratorium on the movement of bees and beekeeping equipment to try to get an idea of the mite's distribution, but its presence had already been confirmed in Pennsylvania, Ohio, Illinois and Wisconsin.

"Emergency teams made up of Florida bee inspectors and the US Agricultural Department's Aphis [animal and plant health inspection service] are now combing the state for infected colonies," Sanford reported. The scale of the crisis was demonstrated by Canada and Mexico sealing their borders to US bees.

Today, there isn't one state in America that doesn't have varroa, and, as in Europe, beekeepers have been forced to make changes on a scale not seen since apiarists moved from keeping bees in skeps (straw baskets) to moveable-frame box hives more than 150 years earlier. Sanford goes so far as to characterise two types of beekeepers: before varroa (BV), when theirs was a laissez-faire, bee-focused pastime; and after varroa (AV), when it became about keeping one step ahead of the parasite.

By 2005, there were about a third fewer honeybee colonies in the States than before the mite's arrival, yet when US scientists began investigating the causes of CCD, they quickly removed varroa from the list of prime suspects. The reason they gave was that there was little evidence of the mites in some of the samples of bees from CCD-infected hives. What they seemed unaware of were ominous findings some years earlier that showed that although large numbers of mites were required to produce an initial viral infection, once established the virus could endure without the mite.

Norman Carreck, a former apiculturist at the UK's Rothamsted research station, recalls how controlled studies at the centre showed that varroa could transmit Kashmir bee virus (KBV), a relation of APV that killed bees in two to three days. Moreover, the virus could persist in the absence of mites, suggesting that bee-to-bee transmission can take place.

"This should not surprise us," says Carreck, who points to varroa-free Australia – where KBV is common in bees – as evidence that there must be a natural means of transmission from bee to bee. What that means of transmission is, no one is yet sure, but it could prove a vital missing piece in the CCD jigsaw.

Bees don't go around sneezing, so they couldn't pick up a virus from their sisters in that way. But they do exchange saliva when house bees receive nectar from foragers, and pellets of pollen are moistened with saliva when they are packed into cells before being eaten by other bees. New research from Cox-Foster and her colleague, Nancy Ostinguy, identifies for the first time pollen as a viral transmission route.

"We found that the pollen on the flowers can get contaminated with these viruses," says a clearly excited Cox-Foster. "I'll try not to sound grandiose here, but if you do have a virus introduced into an area, such as California where half the US bees go for pollination, there is the chance for it to move very broadly if it's in the food."

Contaminated pollen could also explain why other bees and pests such as wax moth and small hive beetle shun CCD-affected hives rather than rob them of food, as normally happens when colonies die.

European and UK varroa experts, however, suspect that miticide-resistant mites are at the heart of today's carnage. After enduring 20 years of chemical warfare, the mites have built up a natural resistance to what ever is thrown at them.

Pyrethroids are manufactured chemicals very similar in molecular structure to pyrethrins, which occur naturally in some chrysanthemum flowers. They are toxic to insects and mammals and last for a long time in the environment. They kill by prolonging the opening of a cell's sodium channels, which leads to paralysis and death. Animals can become resistant by improving their detoxification by means of enzymes, or changing the shape of their sodium channel so they become less sensitive to the chemical. Either way, once-lethal doses of pyrethroids no longer kill the bees' enemy. For example, fluvalinate, the active ingredient in varroa-control product Apistan, is now redundant in many countries.

There are various theories about why this is so, notably that beekeepers who leave varroa-control treatments in hives for longer or shorter periods than recommended, or use home-made recipes of fluvalinate, have unwittingly selected for resistant mite populations.

In 1991, only four years after pyrethroid use was permitted in Italy, resistant mites were detected in the north-west region of Lombardy, in colonies closely connected to Sicily, where similar problems were occurring. Just like the original *Varroa destructor* mite, the resistant variety was quickly spread by man to neighbouring Switzerland, Slovenia and southern France, where farmers were starting to use the pesticide imidacloprid on their sunflowers. The resistant mites reached Germany in 1997, Finland in 1998 and the UK in 2001. Devon, where varroa had entered Britain a decade earlier, was home to 25 apiaries with the first fluvalinate-resistant varroa. A commercial beekeeper was blamed for leaving Apistan strips too long in his hives. By June 2004, the resistant mites spanned 140 apiaries across the British Isles, some of which were hundreds of kilometres away. The pattern of spread was twofold: a slow fanning-out by flying bees moving between colonies locally; and irregular long-distance jumps caused by beekeepers moving infested hives.

But it wasn't just pyrethroids that the mites had become impervious to. In various countries, including Switzerland,

France and the US, harder chemicals, such as the organophosphate coumaphos, found in Perizin and Check-Mite, or amitraz, the active ingredient in Apivar, no longer worked.

Martin says the time lag between the appearance of resistant mites and new and effective treatments provides the mites with the opportunity to rapidly increase their populations, which can lead to a new virus epidemic. "Because the viral landscape in which it operates is much wider and the viral prevalence much higher than before the spread of the mite, the chances of the mite encountering a virulent strain and passing it on are much more likely, resulting in more rapid death of colonies," he explains.

So CCD in the States, he argues, could be a result of miticide-resistant varroa interacting with a particularly virulent honeybee virus. As a result, only a couple of thousand mites in autumn could kill a colony.

So great was the threat from increasingly impervious mites across Europe that in 1998 joint action was taken by member states of the European Union to promote the development of methods that minimise the use of traditional chemicals in beehives. This was followed up two years later with the launch of a European working group for integrated varroa control.

Peter Rosenkranz, who chaired the working group, is today coordinating a monitoring project on bee losses across

Germany. The project was set up in a climate of suspicion against systemic pesticides after a third of the country's honeybees were wiped out in the winter of 2002-3. Some 125 beekeepers volunteered their bees for the study. Three years and thousands of analysed bee, pollen and honey samples later, the only correlation researchers can find between bees and high mortality is varroa mites.

"The dying colonies had 20-30% varroa infestation," Rosenkranz tells us. "The pattern of losses in Germany is similar to that being witnessed in the US, where one beekeeper can lose 80% of his bees but his neighbour can lose none."

Rosenkranz, who heads the State Institute of Apiculture at the University of Hohenheim in Stuttgart, says that everyone familiar with the destructive relationship between varroa mites and their hosts in Europe recognises the symptoms now being seen in the United States. "We've known of it for a hundred years," he says, with slight exaggeration.

"Even with my own private colonies, you see one day that there are 10,000 healthy-looking bees and three weeks later there are just 1,500," Rosenkranz adds.

Treating colonies three times a year to rid them of varroa mites is a time-consuming and not very sexy or hi-tech solution to stamping out CCD. At a time when entomologists have access to expensive, cutting-edge equipment that can rapidly analyse DNA sequences, identify distinct strains of viruses and even locate genes in the bee that are activated as an immune response when bees are exposed to a pathogen, who can blame them for exploring these areas of scientific inquiry? But on more than one occasion beekeepers, state apiarists and even European researchers we spoke to raised questions about whether the scientists really know how to use the tools, or what to look for.

The political and commercial pressure to find a magic bullet for a phenomenon that threatens the US's multi-billion-dollar agricultural industry could be another reason why the mites are being overlooked. How embarrassing to admit that you just let a few mites get the better of you! Much more impressive to unveil a new antiviral bee drug.

The Danes initially suffered little from the massive bee losses that hit much of Europe and now the States. When varroa arrived in Denmark in 1984, the government took the lead in ensuring that beekeepers adopted a uniform and chemical-lite approach to varroa control. Today, 89% of Danish apiarists follow a three-step programme. In the spring, they remove drone cells, where the mites prefer to lay their eggs; in the summer, they place organic formic acid in the hive after the last honey harvest; and at the end of summer, another organic acid, oxalic acid, is trickled over the bees. As a result, 60% of large-scale beekeepers claim that they have "never seen damaged colonies or bees", while 25%

admit to having just "single colonies now and then that have damage due to varroa".

Different approaches on the Channel Islands also demonstrate the benefit of a universal strategy. When the mite struck Jersey, there was a free-for-all with beekeepers trying all variety of chemicals to kill the parasite, but a third of the bees were wiped out. On neighbouring Guernsey, in contrast, the government made it a criminal offence not to treat hives with Apistan strips that it gave away free to beekeepers. Here, very few bees perished.

In the US, where migratory beekeepers seem a law unto themselves when it comes to administering varroa treatment – not one beekeeper we spoke to seemed to follow the same procedure – there seems to be more of an appetite for trying to create varroa-, or even virus-resistant, bees than there is in regulating how to control the lethal mite.

Jeff Pettis, co-ordinator of the USDA's five-year programme to improve honeybee survival, told a meeting of US beekeepers in 2008: "We know that one of the main things you need out there is a cheap varroa control product that doesn't cause residues and is highly effective. Well, using resistant stock doesn't leave residues."

Marla Spivak has been breeding bees for resistance to varroa mites for more than a decade. Her Minnesota Hygienic line of bees, so called because they were developed at the University of Minnesota where she is an apiculture professor, have greater olfactory sensitivity than the average honeybee, which allows them to detect a larva or pupa that is diseased. They will open up the wax-capped cell and start removing the mite-riddled contents before the mother mite has had time to effectively reproduce. Other hygienic behaviour includes fastidious grooming, in which adult bees remove mites from their sisters, damaging the mites in the process. Spivak is now working with commercial bee breeders to help them select for more hygienic bees. It takes four years to stock an apiary with hygienic queens and drones and their offspring. Spivak fears that they may not have that much time.

Other US researchers are experimenting with breeding bees that have the trait known as varroa-sensitive hygiene. But it's not only in the States that disappearing bees have sparked attempts to breed varroa-resistant strains. There are reports of projects as far afield as Great Mercury Island, off New Zealand, and Whitby in Yorkshire. A network of nine leading European honeybee research groups in the areas of pathology, genetics, behaviour and honey quality, called Beeshop, is also trying to breed for tolerance to disease and parasites.

According to scientists at the US government's honey bee breeding, genetics and physiology research unit, hope for their country's honeybees comes from the foothills of south-east Russia. A decade ago, researchers from the unit in

Baton Rouge, Louisiana, led by Thomas Rinderer, scoured Russia's Primorsky Territory in search of bees that had evolved to hold their own against varroa due to heavy mite pressures there. Ten years on, Rinderer says the bees they have bred from those initial Russian queens are capable of fending off not just varroa but three more of the honeybee's worst assailants: tracheal mites, which kill by clogging bees' airways; small hive beetles, which overrun hives and eat brood; and cold temperatures. These bees are less likely to die during harsh winters and appear more frugal with their winter stores.

Rinderer hopes to use the recently sequenced bee genome to pinpoint the genes that make his bees remove mite and beetle-infected brood, groom meticulously, and be thrifty with their honey. Armed with this knowledge, he aims to create a genetically modified superbee from Russian stock. This may sound far-fetched, but the bee genome has already been used by federal scientists to select for resistance to American foulbrood disease.

The danger of creating a superbee, however, is that a superbug would more than likely follow in its wake. And the western honeybee already has enough ordinary foes to contend with.

Small hive beetle, for example. In Africa, where it originated, this scavenger is only a minor pest, as African honeybees have strong house-cleaning and defensive traits that limit its reproduction. The bees deny the beetles access to the hive by aggressively harassing them, fill cavities where the beetle could hide with resin-like propolis, quickly remove beetle larvae, and confine the beetles to "propolis prisons".

Within two years of the beetle's discovery in the US in 1998, however, it had destroyed at least 20,000 colonies. Its arrival in Australia two years later was similarly harmful because western honeybees have no natural defences against it. They allow the beetle to gatecrash and lay her eggs in their home, and her offspring wreck the furniture by tunnelling through wax comb to eat bee brood, pollen and honey. Because a beetle can lay hundreds of eggs, it takes relatively few to overrun a hive and cause the rightful owners to abscond.

But does the small hive beetle play any role in CCD? Although the bees disappear, no evidence of the beetle has been observed in CCD-affected hives. Scientists think, however, that the beetle may carry viruses. One of Ritter's German students is working at the Beltsville bee lab in Maryland to assess its impact.

What about the tracheal mite, *Acarapis woodi*, which can block a bee's windpipe? Does it have a bit part in the drama? It was once thought to have played the lead in the Isle of Wight disease at the beginning of the 20th century. However, the eminent British bee pathologist Lesley Bailey proved years later that the mite was not the main villain. He identified a virus as the bees' killer.

The first 100 samples of CCD-infected bees that were collected from 10 US states and dissected by the CCD working group in the early days of its investigations found "significant levels" of tracheal mite in only one beekeeping operation. They were "not detected, or at low levels" in the remainder. As a result, the windpipe-wounder was dismissed as a prime suspect.

And the parasitic fungus nosema? It is considered one of the most prevalent and damaging of honeybee diseases, says geneticist Robert Paxton of Queen's University in Belfast. But he says it often goes unnoticed because the spore is invisible to the naked eye and the disease rarely leads to the death of a colony.

There are two types of this fungus, both of which impair the digestion of pollen and shorten the life of their hosts. *Nosema apis* has lived in the western honeybees' gut for millions of years, while *Nosema ceranae* is endemic in the eastern honeybee. But in just the past decade *Nosema ceranae* has been accused of jumping over to the western honeybee and spreading rapidly around the world. While the more familiar pathogen makes itself known through dysentery, diarrhoea and crawling bees that are unable to fly, the newer, more vigorous variety is associated with a lack of vigour and fitness, as well as heavier intestinal injuries.

Spanish researchers have channelled their energies into studying the western honeybees' latest foe. As a result, they have become convinced that *Nosema ceranae* is behind the latest die-off in Europe and America. They describe the symptoms as similar to CCD: reduction in the number of bees in a colony, with no apparent cause, until the point of collapse – and no signs of mortality in front of the hives. *Nosema ceranae* was the reported cause of 20,000 colony losses in the Salamanca region of Spain in 2004 and has been accused of causing massive colony losses across the country over the winter of 2005-6.

Mariano Higes, head of the bee pathology laboratory at the regional apiculture centre in Marchamalo, central Spain, has been closely studying the impact of *Nosema ceranae* since May 25 2005. On that day, his team of scientists injected 255 hives in their apiary with the spores of the fungal parasite. Eighteen months later the hives had all dwindled in number and then collapsed. But until that fateful November, the bees seemed to be functioning normally and looked strong and healthy, Higes told a meeting of US beekeepers in 2008.

However, close examination of samples of bees taken from the hives revealed a high number of foragers and nurse bees infected with *Nosema ceranae* just before collapse. Additional hives located near the infected colonies to see if the infection would spread also tested positive for *Nosema ceranae* spores in more than half of their foragers, as did several bees found dead 750 metres away from the hive.

"This is why we really believe that the fungal parasite is causing big bee losses," Higes told his audience.

According to Higes, all the hives in the experiment were free of pesticides and varroa. Yet a number of viruses transmitted by the varroa mite were detected in the samples, including DWV, CBP (chronic bee paralysis) and, in one case, IAPV. So could it be that *Nosema ceranae* can also transmit or amplify viruses? Or is it that bees are less able to fight off a virus if they are debilitated by severe stomach ache?

US scientists quickly crossed the gut parasite off their list of CCD suspects for the simple reason that they found it in both CCD-affected and healthy bees. But were the bees they thought were in fine fettle really as robust as they appeared? According to the Spanish studies, if the bees had *Nosema ceranae* it indicated that the colony was not in good shape and would be dead 18 months later.

Paxton shares some of the Americans' doubts: "There is a correlation between *Nosema ceranae* and colony mortality, but this does not mean that *Nosema ceranae* was the cause of the colony mortality," he says. "There may, however, be a synergistic relationship between *Nosema ceranae* and other factors leading to increased colony mortality."

In the absence of an answer to CCD, US scientists are starting to think that because the parasitic fungus can knock out a part of the bees' immune system, it may well be a player, acting in concert with other pathogens or conditions.

Likewise, Americans seem to be coming around to the idea that varroa has a central role to play in the tragedy. A monthly newsletter written by the respected Californian bee broker Joe Traynor has drawn readers' attention to Martin and Carreck's research in the UK.

"A current theory making the rounds, and gaining acceptance, is that a virus or a combination of viruses is the culprit," wrote Traynor. "The virus, or viruses, is rapidly spread by varroa in the summer and by the time the fall varroa treatment is made, the virus is already well established. The bees still look great going into the fall, but clusters dwindle in the winter and finally collapse to a handful of bees and a queen in January."

Beekeepers, in their strong-held belief that pesticides are to blame for honeybees' demise, admit that the crippling effect of gut ache or a blood-sucking parasite could make an organism more susceptible to a pesticide, or vice versa that exposure to a pesticide makes an organism more susceptible to the diseases caused by *Nosema ceranae* or varroa.

In Germany, Ritter agrees that varroa plays a role in spreading viruses, but cautions against pinning the blame on the mite alone. "We don't know the whole story," he says.

What happened in the first year of the CCD investigation was that each scientific community in the race to identify the honeybee killer concentrated on the suspect they knew most about. The northern Europeans, priding themselves on their varroa expertise, were gunning for the blood-sucking parasite. The Spanish, now a world authority on *Nosema ceranae*, pushed the fungus's credentials. The Americans stood firm behind the IAPV virus, whose identification they claimed was the first major breakthrough in the case. With egos, reputations and careers on the line, not to mention future funding, the stakes were high. As the investigation entered its second year, with no clear winner in view, there were signs that the Americans' gung-ho, go-it-alone ethos was giving way to a more conciliatory approach, that closer cooperation was beginning to take place with the Europeans in a bid to track down the honeybees' assailant. But without a global pool of research money to draw on, separate leads will continue to be pursued.

Whatever tools the scientists use, and whatever tip-offs they follow, however, they will no doubt arrive at a similar conclusion: that there is no single culprit. It is more than likely a combination of parasites, whose armies have grown stronger, and in cohort with other recruits is spreading viruses that the battle-worn honeybee is no longer able to fight off.

One reason for her feebleness is that a major ally has deserted her in her hour of need. By destroying the environment in which the honeybee lives, man is reducing her chances of survival.

CHAPTER 8

THE CHANGING ENVIRONMENT

Bret Adee tells us to meet him on the outskirts of town. There is something we should see. "I'll be in the parking lot of the McDonald's at the intersection where Rosedale Highway meets Allen Road," says Adee, who with his father and brother runs the largest commercial bee operator in the States, Adee Honey. "I'll be in my wife's white car. It should take you about 20 minutes." The journey from downtown to the north-west edges of Bakersfield is illustrative of another problem facing honeybees: low-density urban sprawl.

Downtown Bakersfield has an old-fashioned, almost quaint feel. Two- and three-storey turn-of-the-century houses grace quiet tree-lined streets. There is a main drag with a few small, independent shops, selling trendy clothes, photographic equipment and groceries. On the corner is the art deco Fox picture house, now an entertainment venue, opposite a pedestrianised boulevard, where you'd expect to see people sitting outside on a Saturday afternoon. But the place is deserted. We put down the roof of our hired convertible and head west. Suddenly we see everyone. They are in their cars, sitting in a traffic jam on the freeway that circles town. We turn left onto Stockdale Highway and crawl through what looks like a wealthy business park, with low-rise retail units and insurance and oil company offices adjacent to the Kern City golf course and the Stockdale country club.

Bakersfield's first spurt of growth was at the beginning of the 20th century when the Dust Bowl drought, immortalised in John Steinbeck's Grapes of Wrath, forced thousands of destitute people from Arkansas and Oklahoma to set out for California in search of jobs in the oilfields and cotton planta-tions of Central Valley. The city, which sits at the southern end of the valley, 110 miles (175km) north of Los Angeles, grew steadily over time, with residents mainly employed by the oil industry. As the population rose, south-west Bakersfield sprung up from the sandy loam: a suburb provid-ing near-identical homes, schools, businesses, and leisure and shopping facilities in rabbit-warren-style subdivisions.

At Coffee Road we head north, over the Kern River, where the scenery gives way to less verdant vistas. The blocks become much further apart. We take a left at Rosedale, where drives, streets and avenues of modern suburban homes with freshly laid lawns cover the once-virgin terrain. The further west we head, the wider the blocks, the newer the developments. You can almost smell the newly tarred roads extending their tentacles into the Californian desert to accommodate the latest influx of migrants: young, white middle-class families fleeing LA's crime, smog and sky-high real estate prices.

In 20 years, Bakersfield's population has more than doubled in size to over 250,000. Masterplanned neighbourhoods with names like Riverlakes Ranch, Madison Grove and Brimhall have gobbled up large stretches of desert.

Finally, just as we thought we had overshot the sprouting city, we come to an intersection with a petrol station and a McDonald's amid a cluster of fast food chains. We drive in and look for Bret.

A short, slightly overweight man in his late 30s with thinning blond hair shouts our names. He is wearing jeans, a checked shirt and a leather jacket, and looks more like an ageing college graduate than a beekeeper. He suggests we follow him to his nearby house, where we can pick up his 4x4. We drive a few blocks north, past plots of land on one side where only a few bricks have been laid, and recently constructed ranchettes on the other.

We stop at the end of an affluent looking cul-de-sac full of Grecian-style bungalows. Adee's palatial pile is spread out to the right. His young children are playing with a remote-control aeroplane on the manicured lawn. In the driveway sits the 4x4. Beyond the house is desert. We have come as far as the tentacle stretches for now.

In California, the fruit bowl of the United States, 75,000 acres (30,000 hectares) of orchards, vineyards or fields are paved over each year. The American Farmland Trust says that every single minute of every day, the country loses two acres (four-fifths of a hectare) of agriculture land. During the five years from 1992 to 1997, more than six million acres (2.4 million hectares) of farmland were converted into more profitable houses, malls and roads.

"Each year, you have to drive a little further out to find it. Slowed by traffic, through tangled intersections, past rows of houses that seem to have sprouted from the field, finally you can see the bountiful farmland," the trust laments.

The land grab does not correspond to a comparable increase in the number of people needing new homes. From 1982 to 1997, while the US population grew by 17%, urbanised land expanded 45%. The concrete invasion is a result of the acreage per person for new housing almost doubling in the past two decades.

The urban sprawl makes it that little bit harder for migratory beekeepers to find suitable, accessible sites on which to keep their bees. With no land of their own, they rely on farmers and landowners who are willing to rent.

Adee brought his 70,000 hives to spend the 2007-8 winter in the Californian desert. They are scattered across 4,000 acres (1,600 hectares) of remote scrub owned by three ranches out in the desert, 30 minutes north-west of Bakersfield. The lunar-style landscape is dotted with oil pumps, electricity pylons and clumps of white hives. A few years ago, Adee could have kept his bees closer to the almond orchards they would pollinate in February. But as almond prices rocketed and growers got fat on the profits, more and more ranches with high gates and walled gardens have shot up on land that once housed bees.

The winter digs for John Miller's 13,000 colonies used to be half an hour north-east of Sacramento. As the shopping malls, housing developments and golf courses encroach, it is harder for him to find a spot that provides enough nectar and pollen for the bees to survive until pollination season begins.

Further south, Ron Spears, second only to the Adees in the scale of his beekeeping operation, says that since he started in the business 28 years ago, whole swathes of land around San Diego where he used to keep his bees have been swallowed by real estate speculators.

But it is not just in California that urbanisation is a problem. One beekeeper who divides his time between Maine and Florida said that every winter when he goes south there are more condominiums and less space for bees. Even in the more sparsely populated Vermont, rising house prices over the past decade have led to new developments mushrooming in areas where onions, cucumbers, carrots and other crops once grew.

Competition for scarce water resources between the expanding cities and existing farmland is another consequence of the urban invasion. Lyle Johnston kept thousands of hives on what was once prime farmland in the county of Crowley, Colorado. "There were purple alfalfa blooms for as far as you could see," he recalls. "Now these same locations would not support one hive of bees, with no blooms of any kind for the bees. Today, it is only a county of tumbleweeds and very little farming, if any, is going on."

In the mid-1970s the burgeoning communities of Colorado Springs and Aurora, a suburb of Denver, bought the water rights to the land, knowing that one day they would need the precious commodity to accompany their growth. The irrigation water was diverted in the mountains from what had been some of the best bee pasture in the US, leaving the clover-rich landscape to dry up.

A similar pattern is now emerging across the US, particularly in drought-prone regions like California, where cities such as LA have priority for water over farming. In 1992, the state's water project for the first time suspended all water deliveries to agriculture after a six-year drought left too little in the reservoirs to meet urban needs. As shortfalls become more frequent, farmers are starting to feel the pinch, says Johnston. Moreover, they can make more money selling water rights than tilling the land. By 2010, San Diego will need double the water it used in 1994. "The cities are willing to pay whatever it takes to secure water for the future, and farmers can't afford to keep the water to farm," Johnston says. "In Colorado most of the farmers were ready to retire so they took the money for the water and ran to the bank."

Although he still keeps bees in Colorado, the area is much smaller and he has halved his operation from 4,000 to 2,000 hives.

Fewer green spaces means fewer wild flowers and plants for bees to forage on and sustain a healthy diet. Unfortunately, many fabulous-looking cultivated blooms, such as double-headed roses, chrysanthemums and dahlias, provide no nectar and little pollen. And suburban gardens fashionably covered in easy-to-maintain decking or landscaped with exotic succulents and tropical palm trees are a wasteland for honeybees.

Urban sprawl also inevitably brings humans and bees into closer proximity. With sensationalist headlines screaming about the spread of Africanised "killer" bees across the States, people are increasingly uncomfortable about sharing space with bees, so there is a danger of beekeeping being forced out of towns. Battles in the suburbs between long-established beekeepers and newly arrived families with children are a growing problem – so much so that many commercial beekeepers in the mid-west and northern US now store their hives indoors, in cool, dark potato cellars in Idaho from November to late January, rather than trucking them to warmer climes down south.

With 16 hives located on rooftops across New York, from the Upper West Side, to Harlem, downtown on a 12-storey hotel in the East Village and over in Brooklyn, David Graves is one of the most urban beekeepers in the world. He is also one of those most wedded to the theory that mobile phones are the cause of CCD.

We meet him by accident at a green flea market – which sells a mixture of local organic food and secondhand clothes – on the corner of Columbus and 77th on a cold, sunny morning a couple of blocks west from a snow-covered Central Park. He is selling his Rooftop brand honey and telling anyone who cares to listen that radiation from cellphones and the masts that dot towns and cities are killing the bees.

"I believe these towers are messing up my bees' navigation and their ability to communicate and find their way home," he tells us.

Last year all his bee colonies died, and again this year he says he will have to restock the hives. He says his bees are especially vulnerable to cellphone towers because they share the rooftops with them. When we tell him that the original newspaper article reporting a study linking mobile phones to CCD, which was picked up by the world's media as a possible explanation for vanishing bees, had got it wrong, he didn't want to know. He had just fought a successful campaign in Brooklyn to stop a cellphone mast being located near one of his hives. "You can say what you like, but those towers are damaging my bees," replies our curmudgeonly honey-seller.

The Landau University study that first appeared in Britain's Independent newspaper was looking for effects of radio frequencies on honeybees and suggested that when cordless-phone base stations – rather than the misreported mobile phones – are put in beehives, the close-range electro-magnetic field may reduce the ability of the bees to return home. In the course of the study, half of the colonies collapsed, including some from a control group.

Stefan Kimmel, a graduate student who co-authored the Landau study, said of the newspaper article: "This is a horror story for every researcher to have your study reduced to this. Ever since the Independent wrote their article, none of us have been able to do any of our work because all our time has been spent in phone calls and emails trying to set things straight." He added: "If the Americans are looking for an explanation for colony collapse disorder, perhaps they should look at herbicides, pesticides and especially geneti-cally modified crops."

New York's filthy air could be what's troubling Graves's bees, judging by a study from the University of Virginia. Air pollution in cities is destroying the smell of flowers before it has a chance to spread, making it difficult for bees to find their food source, researchers found.

Jose Fuentes, who led the study, published in the journal Atmospheric Environment in April 2008, said: "The scent molecules produced by flowers in a less polluted environment, such as in the 1800s, could travel for roughly 1,000 to 1,200 metres [1,100-1,300 yards]. But in today's polluted environment downwind of big cities, they may travel only 200 to 300 metres [220-330 yards]. This makes it increasingly difficult for pollinators to locate the flowers."

To test out his idea, he created a mathematical model of how scent molecules travel on the wind. The scent molecules from flowers are volatile and easily bond with chemical pollutants from car exhausts and chimneys, such as ozone, hydroxyl and nitrates. These reactions destroy the fragrances so that, instead of travelling long distances on the wind, the scent is neutralised close to its source. "It quickly became apparent that air pollution destroys the aroma of flowers by as much as 90% [compared with] periods before automobiles and heavy industry."

Although beekeepers have not reported seeing less food coming into the hives when the weather is good, Fuentes fears that the fading smell of flowers may stress insects already faced with an array of other threats. "The [effects shown in] these studies will simply exacerbate whatever the bees are going through right now," he said. "It's something that is really worthwhile paying attention to."

Away from the big smog, North and South Dakota still provide good summer lodgings for millions of honeybees. Pastures are strewed with white clover in the spring, followed by purple alfalfa and pale pink buckwheat. Fading prairie towns, sparsely populated by ageing farmers, are glad of the rental income the beekeepers provide. But changing farming practices mean there are fewer and fewer such places in the US where bees can feast on wholesome food. Huge fields of grass are mown and dried to make hay for cattle before clover has had a chance to bloom. And increasing demand for animal feed has led to large tracts of land planted with alfalfa being cut just before bloom. Moreover, the rise in the price of maize, driven by growing demand for ethanol for biofuel, has meant that beekeepers in the north and mid-west are now surrounded by a sea of sweetcorn as farmers transform fields of less profitable crops. And while the money's in corn, farmers are less likely to rotate it with more bee-friendly plants. Bees collect pollen from corn in the absence of anything else, but it fails to provide them with a balanced diet, and yields no energy-boosting nectar.

"One of the biggest problems that will face beekeepers in the future is the lack of bee pasture to keep bees healthy," says Johnston. "There is no chance at all of keeping strong hives for the almonds [pollination in California in February] without a decent summer honey crop."

The trend to convert swathes of arable land to biofuel production has swept throughout the globe from China to India and Australia, raising questions about how to feed the world when staple crops are now being grown to power cars. More and more land in the US and Europe is expected to make the switch. In Europe, bright yellow fields of rapeseed oil and sunflowers testify to the change. In 2004, the EU harvested 7.5 million hectares (19 million acres) of oilseeds for biodiesel, more than half of it in Germany, followed by France and Italy. While these crops do at least allow bees to produce copious amounts of honey, like humans, bees don't fare well living off just one type of food.

The behemoth scale of biofuel production is merely the latest manifestation of an industrialised system of crop cultivation that is the hallmark of modern agriculture. The world's 1.5 billion hectares (3.7 billion acres) of arable land is now blanketed with single-crop plantations and orchards that stretch for hundreds of kilometres. Apples in China, soya beans in Brazil, almonds in America and every one of the world's other 115 commercial crops are grown in this fashion. The advantages for the farmer are in economies of scale: the crops bloom at the same time, use the same fertilisers and pesticides and can be harvested together for maximum efficiency.

Most oil seeds, fruits, nuts and vegetables that we eat depend on pollination by animals such as butterflies, bats, and bees. But this intensive method of production robs pollinators of diverse food supplies and places to set up home. Weeds and hedgerows on the edge of fields could compensate, but they take up valuable space in the brave new world of mechanised food production. As a result, native pollinators all over the world are in decline. Some 1,000 solitary bee species alone are on the verge of extinction.

Before honeybees were brought to North America from Europe in the 17th century, over 4,000 native species of bees pollinated the continent's fruits, vegetables, weeds and trees. One of the advantages of the newcomers was that honeybees can be housed in hives, looked after and manipulated, and transported en masse by skilled apiarists. Some even argue they are farmed. This has led growers to become more and more dependent on honeybees to do the job once performed by a myriad of different species. Their profits now hinge as much on honeybees' availability – sometimes for as little as a few days – to pollinate their blossoming fields, as they do in the sun and the rain.

But honeybees don't just provide a valuable service producing fruits and vegetables for the kitchen. They maintain the biodiversity of flowering plants, whose seeds and fruits feed wild birds and animals and form the basis of almost all terrestrial food chains. They also enable many soil-enriching and soil-holding species of wild plants to reproduce.

Cranberries and blueberries are particularly poor sources of nutrition for honeybees, yet without frequent visits from nature's master pollinator at the right time of year there would be up to a 90% reduction in the yield of berries on the bushes. A single-course meal for hundreds of kilometres forces migratory beekeepers to truck their bees to their next pollination appointment as soon as the job is done, or to find increasingly elusive bee pasture with a choice of dishes.

Eric Mussen, cooperative extension apiculturist at University of California, Davis, believes malnutrition is a key factor in colony collapse disorder. Malnourished bees are more susceptible to disease, predators and insecticides, he points out, while the best-fed bees are the hardiest. He says a mix of high-quality pollens in August and September is the key to keeping bees healthy through the winter.

But Mussen is more concerned with changing weather patterns, rather than farming practices, damaging honey-bees' nutrition. The climate can affect bees' food supply in a number of ways: excess rain keeps bees indoors, preventing them from visiting flowers for food; a lack of rain, or too much heat, reduces the amount of nectar and pollen a plant produces; and cold nights can create sterile pollens that don't contain sufficient nourishment.

"If we're headed into rougher weather, as it appears we are, we'll have more difficulties with our bees," he predicts. "It won't matter if you're a backyard beekeeper or someone with 10,000 colonies."

In the summer of 2006, many regions in the US experienced severe drought, leading to a total honey crop of just 155 million pounds (70 million kilos), one of the lowest on record. In California the crop was down 30%, and in North and South Dakota honey production fell 15% and 40% respectively. Miller remembers the unprecedented heat wave in North Dakota in July that made the nectar give out early. The alfalfa was so dry and bitter that he harvested the little honey early and got his bees ready for winter ahead of schedule, feeding them to augment their pitiful winter stores.

Beekeepers try to bolster their bees' diet with artificial pollen supplements rich in protein, vitamins and minerals with names such as Bee Pro, MegaBee and Feed Bee. But most accept that these often soya-based patties, liquid or powders, will never replace nature's natural larder.

The year 2007 was the driest on record in California, where 2in (5cm) of rain fell instead of the usual 17in (43cm). Spears says he had no choice but to pull all his bees out of the state and truck them north: drastic action he had never had to take before. "There was no grass, nothing in March and April, no chance of rain, so nothing to stay for," he recalls. "Usually half my bees are making sage and buckwheat honey in the foothills of Saint Gabriel's mountains [north-east of LA]; others are making honey blossom in the foothills of Sierra Nevada. But there was nothing last year, so we moved them up to North Dakota and Montana."

It was a similar story of extreme weather across the globe. Droughts hit eastern Europe, while northern Europe had too much rain. And when summer arrived in the southern hemisphere it was accompanied by disastrous beekeeping conditions.

Drought in Australia killed off plants in major honey-producing areas. In Argentina, the world's biggest exporter of honey, many parts of the country had no rain for weeks, followed by heavy downpours and even flooding.

Autumn storms seem to be a seasonal hazard in the Pampas region of Argentina, home to the capital Buenos Aires and the cities of Cordoba, Rosario and Sante Fe, as well as millions of square kilometres of fertile plains for grazing, agriculture and honey production. In March 2007 – autumn in the southern hemisphere – huge swathes of farmland were submerged under water and 20,000 people evacuated from their homes as rivers burst their banks after a week-long rainstorm. (In 2001 and 2003, there had been similar scenes as thousands of hectares of crops were ruined and people displaced after the region was battered by storms.)

The coldest winter since 1961 followed the deluge, with snow in most parts of the country in early July. And spring provided little respite, arriving with two weeks of torrential rain and continuing cold and wind.

The inclement conditions were a recipe for disaster. As well as reducing the amount of nectar collected, the summer drought had limited supplies of high-quality pollen. So the bees went into autumn hungry and malnourished, only to find that their keepers couldn't get out in the water-logged terrain to feed them supplements and treat against blood-sucking mites and fungal disease. When winter arrived, the bees were ill equipped to fend off parasites and pathogens, and their meagre honey stores quickly ran out. By the spring, hundreds of thousands of colonies had starved to death.

According to Argentina's ministry of agriculture, 30% of the country's 3.7 million honeybee colonies were wiped out in the past two years – more than one million hives. Such a high mortality rate is a devastating blow to a sector now worth $136 million (£97 million), the bulk of which comes from honey exports. In 2007, Argentina – which overtook China as the world's number one honey exporter five years ago – produced 73,000 tonnes (160 million pounds) of honey, compared with 100,000 tonnes (220 million pounds) the previous year.

This downturn has precipitated a "serious economic crisis" in the honeybee industry, according to the agriculture ministry. The financial ruin faced by many of the country's 9,000 or so commercial beekeepers was another reason why hives went without basic varroa control treatment and nutritional supplements.

The government is working with members of the National Beekeeping Council to finalise a plan to rescue the industry. But it is not just adverse weather that threatens Argentina's honeybees. The vast savannas that have supported herds of dairy cows and beef cattle, and bee-friendly pastures, are fast disappearing as farmers swap livestock for soya beans.

"The problem is that farmers are not making intensive grassland management compatible with the honeybee," says Martin Braunstein, the country's largest queen breeder and a member of the National Beekeeping Council.

"Much of the crisis in the beekeeping industry is due not only to the overwhelming expansion of soya beans," he adds, "but to the collapse in the number of farmers, who increasingly have less incentive to continue producing meat and milk and keep alive the pastures that feed our bees." Asked if beekeeping is sustainable in these conditions, he replies, "Definitely not. Except for those [honey] producers who have it as a sideline and have some income from elsewhere."

*

One backyard experiment suggests it's not just greenhouse gases that are altering the climate as far as beekeepers are concerned. Wayne Esaias, a Nasa biological oceanographer who keeps bees as a hobby, blames urbanisation for temperature rises in suburban Maryland.

For the past 15 years, on spring and summer nights, he – or his children before they went off to college – religiously weighed the 17 hives in their garden in the community of Highland. One day, out of curiosity, he plotted the figures on a graph. He tells us what he saw was "tremendous variables from one year to the next, but also a disturbing long-term trend of the hives getting heavier earlier and earlier in the year". What that meant was that the bees were collecting nectar and pollen earlier because the trees that were their primary source of nectar in the neighbourhood were flowering earlier than before.

Once he had observed this trend, he searched for other local data sources to corroborate what his bees were telling him. He found one record dating back to 1922-3 in what used to be a nearby USDA agricultural research centre, and another survey done in 1957 by a researcher at the University of Maryland. Following adjustments for differences in elevation, the older records confirmed his own findings: that there had been dramatic advances in peak flowering times.

Esaias discovered that a botanist at the Smithsonian Institute in Washington DC had kept track of flowering dates for trees in and around the US capital since 1970. This data was based on city residents observing when trees in their garden were in bloom. The trees included tulip poplars and black locusts, two of Esaias's honeybees' favourite nectar sources. Comparing the tree surveys with Esaias's own hive-weight, graphs showed that the bloom period in Maryland was now a whole month earlier than in 1970.

Pavements and roads, less soil moisture, air pollution, and heat from homes and offices all conspire to raise city temperatures compared with surrounding rural areas. So, as cities expand, so too does their heating effect, and flowers' blooming patterns change. In Maryland, there has been a 4F (2.2C) rise in the average temperature for the first three months of the year since Esaias began beekeeping.

"There is a good bit more development around here in the last 15 years," he says. "There is a definite urbanisation component to this change in temperature versus general climate change."

But is it a problem for honeybees if our springs get earlier? Flowering plants and pollinators co-evolved over millions of years, so won't honeybees just adapt to the new conditions? Worryingly, there are reports that trees such as red maples and pussy-willows are blossoming weeks before it is warm enough for the bees to fly in spring.

"There is no evidence to prove that trees and bees have different thermometers. They could, however, have different prompts," warns Esaias. "No one really knows what the relationships are between weather, pollinators and plants. So things could be out of kilter."

A cold snap during a warm period could exert a different stress on a colony of bees than on a tree, he points out. Whereas trees may not feel the cold because their roots are well insulated, a very cold spring night could kill off newly laid eggs around the edges of the cluster of bees in a hive.

So could climate change be a cause of CCD? After all, reports of bee populations dwindling and disappearing usually follow a cold spell. Esaias thinks it could contribute to the stresses on colonies, but in more subtle ways than bees freezing to death. "Maybe a cold spell means different pollen is available that is not so nutritious," he suggests.

Esaias is aiming to set up a national network of "hive hefters" who will act as volunteer climate change monitors. By collecting similar data at sites across the US, he hopes the results could offer clues about how to better manage bee colonies and keep them healthy as temperatures change. For example, if flowers in an area are found to be blooming early in the year before honeybee colonies are populous enough to maximise nectar and pollen supplies, beekeepers would know they had to build up their bees as winter ends.

But it's not all about earlier springs. Just as climate change does not mean universal rises in temperature – Britain, for example, could lose the Gulf Stream, which brings with it mild winters – some plants are flowering later. "If plants don't get cold enough to rest over winter, then blooming can be delayed," Esaias explains. "Sweet clover in Washington now blooms later and the nectar flow is later in Georgia."

His project could have major implications for migratory beekeepers who truck their bees south every winter in the belief that the warm climes create a feeding frenzy, when in fact the nectar and pollen could be on tap weeks later than they were 30 years ago.

In 2007, Esaias enlisted the help of 14 other beekeepers in Washington, Delaware and Maryland, from Harrington to Great Falls. Initial results from recording the daily weight of their hives showed a 15-day gap between nectar production in the built-up neighbourhood of Chevy Chase and the more rural Highland district, where Esaias lives, 20 miles (32km) away. Since the former is also about 200ft (60 metres) lower, Esaias has worked out that 10 days of the gap are attributable to elevation and five days to urbanisation.

In 2008, with funding from his employers at Nasa, he has recruited backyard beekeepers in 68 sites, ranging from Maryland to Virginia, Ohio and Texas. Because bees forage over an area – 2,500 acres (1,000 hectares) – that can be easily mapped onto computer models used to predict ecosystems' response to climate change, the hive weights can help to build a more sophisticated picture of how plants and animals will respond to climate change.

The results will also be used to chart the potential spread of the Africanised honeybee across the States. Although the transportation of western honeybees across continents is without a doubt the main reason that they have come into contact with so many pests and parasites, there are fears that one of the main threats posed by climate change is that it will facilitate the spread of more exotic maladies.

The sting-happy African honeybee (*Apis scutellata*) was introduced to Brazil in the early 1950s, and soon cross-bred with its more gentle western cousin. The spread of their offspring, the Africanised honeybee or AHB, from South to North America has been described as "one of history's most spectacular examples of biological invasion".

First detected in Hidalgo, Texas, in 1990, the Africanised bee has colonised the American south-west and parts of California, and two years ago moved east to Florida. Some climate models predict that it will never reach Maryland, others that temperatures will rise high enough for it to get almost as far as Canada, changing beekeeping in the US for ever.

The small hive beetle is another recent African arrival to the States. It infests hives, and the beetle larvae feed on comb and honey, forcing bees to leave home. It is also thought to transmit viruses. When it eventually crosses the Atlantic in an easterly direction, carried on container shipping or airfreight, it will most likely ransack apiaries in milder parts of Europe. But it has survived as far north as Manitoba in Canada by snuggling up in the hive with the bees during winter.

The parasite brood mite, tropilaelaps, was first reported in 1982 as a new species of parasite on the giant honeybee (*Apis dorsata*) in Sri Lanka. Like *Varroa destructor*, it has jumped species and now lives on western honeybees in Asia and the Middle East. As with varroa, the new host is ill-equipped to deal with this Asian freeloader. Reddish-brown in colour and about 1mm (¼in) long, tropilaelaps resembles varroa and causes similar damage, spreading viruses that stunt adults, deform wings and shorten lifespans. It has claimed the lives of millions of western honeybees in Pakistan and Afghanistan.

The main difference between the two parasites is that tropilaelaps, unable to pierce the membranes of adult bees, feeds instead on the blood of developing bees. This means that it is unlikely to survive in hives where the queen stops laying during winter. But as areas of the world become warmer, honeybee colonies will be continually rearing brood all year round, increasing the potential for the mite's spread.

British entomologists, meanwhile, fear that it is just a matter of time before another bee enemy, the Asian hornet, *Vespa velutina*, crosses the English Channel from France. Global warming has been blamed for the survival and spread of the 46mm (1.8in) insect, which is thought to have arrived in France from Asia in a package of Chinese pottery in 2004. The hornets feed their young with the larvae of bees, whose nests they break into and ransack. A handful can destroy a hive of 30,000 bees in just a couple of hours. Thousands of spherical hornet nests now dot forests in south-west France. "Their spread across French territory has been like lightning," said Jean Haxaire, the entomologist who identified the new arrival. So concerned are British beekeepers that the giant hornet was cited in a strategy unveiled by the UK government in April 2008 to save the country's 250,000 colonies. Measures include tighter controls at ports and a network of "sentinel" hives near ports and airports to alert experts to the arrival of any new pest or disease.

Although these exotic pests do pose major threats to the future survival of the western honeybee species as temperatures rise, many have not yet reached the US and cannot therefore be blamed for CCD. Warmer winters, however, are already allowing honeybees' public enemy number one to grow and spread earlier than usual, with potentially devastating consequences.

German beekeepers were put on red alert in the summer of 2007 when it was discovered that the relatively warm winter had allowed varroa populations to expand to dangerously high levels. There were warnings to treat their colonies much earlier in the year, amid conservative estimates by the bee research institutions that around 30% of the country's colonies could be wiped out. According to Peter Rosenkranz, chair of the German bee losses monitoring group, that is just what happened despite attempts at early intervention.

Ron Spears in California feels his problems are related not to climate change, but to the fact he has to take his bees from cold to warm climates at the onset of winter. As a result his queens are non-stop egg-laying machines and the bees are out foraging most days of the year. "They never have a chance to rest," says Spears. "Ideally I'd leave them in one location, but it's not profitable."

This unnatural cycle is fuelled by the main event in the commercial beekeepers' calendar, the Californian almond pollination in the middle of February. To prepare honeybees for the most valuable three weeks of the year – when they can earn their owners $140 a hive – requires a lengthy build-up ahead of spring. Imagine running a race at the Olympics without months of training. That's what it would be like if beekeepers didn't try to get their bees into tip-top condition for the pollination championship. Their whole year is now spent in preparation for the almond orchards, including the winter when western honeybees naturally take a breather.

But as the bees develop earlier and the colony's population expands, so do their parasites: the mites that may hold the key to their downfall.

Bret Adee certainly thinks the blood-sucking varroa mite played a leading role in the massacre he shows us out in the Californian desert, where 40,000 of his white hives lie empty and silent, scattered across the wilderness like gravestones in a cemetery.

CHAPTER 9

THE INDUSTRIALISATION OF POLLINATION

Joe Traynor was the first person we visited in Bakersfield, California. He is the most prominent bee broker in Central Valley, so we'd called him to learn how the pollination of California's 60 million almond trees could be a major factor in the honeybees' demise.

A small, second-floor apartment in a quiet side street is Traynor's headquarters for six weeks every year. Under the auspices of Scientific Ag Co, the company he founded in 1973, he orchestrates part of the largest, and most profitable, managed pollination event on the planet.

Central Valley stretches from Bakersfield, two hours' drive north of Los Angeles, to Red Bluff some 400 miles north. More than 600,000 acres of it are planted with almond trees. No other place on earth has the soil, water, climate and space to grow the nuts on such a vast scale. Some 80% of the world's almond production takes place here, and almonds have become the state's most valuable agricultural export. In 2007 the record-breaking 1.3 billion pound (590 million kg) harvest earned the state of California $1.9 billion in exports.

It could be argued that the expansion of California's almond industry from 90,000 acres in the 1950s to more than six times that by 2008 is the single most important event in US beekeeping in the past 60 years.

Pollinating this vast string of orchards is a 22-day job that needs some 40 billion bees. Half of all the honeybees in the United States have to be trucked into Central Valley during February. Around 3,000 trucks make the journey from as far as Florida, each carrying some 480 beehives stacked four high. In the cool hours before sunrise more than one million colonies of bees are unloaded between the endless rows of almond trees.

By February 16, the US's national Almond Day, the trees are usually in full bloom and humming with the sound of busy bees, the valley blanketed in pale pink and white blossoms. But the bloom is late when we visit and most trees are ugly and bare. Yet even in the few spots where there is a burst of lavish flowering this is no rural idyll. The sandy loam and Mediterranean climate are perfect for the cultivation of almonds, but that's where any comparisons with picturesque orchards of Spain or Italy end. There are no verdant weeds, wild flowers or grass verges to please the eye, just row upon row of trees laid out in uniform diamond grids.

An acre of almond orchard in this part of the world produces some 3,000lb (1360kg) of nuts – six times more than 30 years ago. The inexorable increase is partly down to new varieties of high-yielding trees and the use of some of the most technologically advanced agricultural methods to tend and harvest the nuts. But more than anything it is due to the planting of varieties that bloom together and maximise cross-pollination by the honeybee.

Because almonds require more pollination than any other tree crop to get a bumper harvest, two hives are recommended per acre. Traynor is responsible for getting 35,000 beehives placed in the southern end of Central Valley. Driving along the freeways in February you can clearly see clumps of white-painted boxes on the edge of the orchards. Some carry labels warning they are fitted with microchips; others offer a $1,000 (£715) reward for information leading to the arrest and conviction of thieves. Stolen hives can be a problem if there aren't enough bees to go round. And this year, with the threat of colony collapse disorder hanging over the almond orchards, there are fears that bees could be thin on the ground. Traynor says you should never go by the number of hives that are spread across an orchard, but by what's in them. Hive strength is measured in the number of frames in the hive that are covered in bees. Traynor guarantees all his beekeepers will deliver "eight-frame" hives.

Traynor matches 40 beekeepers from all over the States with 40 Californian almond growers. He is described by other brokers as "the engine that pulls the train". So it is a surprise to find a modest, softly spoken, grey-haired man sitting at a kitchen table cluttered with unwashed breakfast dishes. He is about to eat the lunch his wife delivers every day while he is holed up in his office from the middle of January. Since moving bees and dealing with clients is a 24/7 job during this time of year, he found it was easier on the family to move out of his home.

The office space is littered with books, papers and magazines. Filing cabinets are bursting with reports and newspaper cuttings. "I'm an information junkie. I stick every article about bees in the cabinets," he explains. The tatty walls are adorned with scenic posters and photos of bees. Through an open door, another shabby room contains an unmade bed. Traynor, who easily looks 10 years younger than his 71 years, has been up since 4.30am. There was a delivery of bees in the early hours and occasionally beekeepers get stuck in orchards or can't find their way because of fog. A usual day includes driving around to check on local bee deliveries when it's still dark. On Saturdays the phone rings less, so he's able to drive out to a 100-mile (160km) radius. He swims three times a week at 5.30am at the local college pool, and can sometimes be found just before dawn having breakfast and talking shop with out-of-state beekeepers in the diner next to the Best Western hotel where many of them stay. From 7am onwards he is on the phone until late at night coordinating deliveries. If a truck breaks down or a beekeeper finds his bees are suddenly dead, Traynor has to find a replacement within hours. The growers won't pay if their bees are late.

While we are at his office two mobile phones ring. A beekeeper has mislaid his map; a grower is checking if Traynor's bee inspector is coming out in the morning.

In the past few years, with the plethora of maladies afflicting the honeybee and the high price of hive rental, almond growers have insisted they get their money's worth.

Richard Emms, who farms 360 acres of almonds near Bakersfield, has no complaints. He has been renting bees from Traynor for seven years. This year he has 320 colonies for 180 fruit-bearing acres. A tall, imposing figure, he dons a veil and accompanies Traynor's inspector on a tour of the hives. He stands close by while the hives are opened and confirmed to be bursting with bees. The Emms land has been in the family since his grandfather moved to California in 1918 and set up a dairy. Cotton, wheat, corn, carrots and alfalfa have all been grown here, but like many farms in Central Valley, it has been converted to raise the crop that pays best.

Farm prices for a pound of almonds tripled from $1 to $3 between 2001 and 2005 and in 2008 were just below a respectable $2 (£1.43). Central Valley's 6,000 almond growers, which range from family-owned businesses with a few hundred acres to large corporations with thousands, have all got rich.

Traynor saw which way the wind was blowing back in the early 1970s. After taking a degree in pomology (fruit cultivation) at the University of California, Davis, and keeping, or "running", as they call it here, 400 beehives, he spotted a gap in the pollination market. Then, there were around four million colonies of honeybees in the United States and hives rented for $3. Today, following the destruction caused by varroa and a collapse in honey prices, only around 2.2 million colonies remain. As almond acreage and bee numbers began to head in different directions, the shortage of bees, even before colony collapse disorder struck, had sent the price of hive rentals soaring. Now, with the price at an all-time high of between $145 to $180 (£104-£129) a hive, beekeepers across the country flock to California, leading to comparisons to the Gold Rush of 1849. For the majority, if not all, of America's 1,000 commercial beekeepers, the three weeks of almond pollination keep their business afloat.

Traynor himself earns $11 for each hive he brokers. He donates $2 a hive to bee research but that still leaves $315,000 (£225,000) – not bad for six weeks' work. He insists, however, that it's a full-time occupation. He spends the rest of the year lining up things for the following season, including visits to bee operations to see if they meet his standards, drawing up contracts and deciding a price for hive rentals months ahead.

Making his job more challenging are fears that because of colony collapse disorder beekeepers won't be able to deliver in February what they promise in July, or even December, when their bees seem strong and healthy.

When Traynor wrote the definitive guide to almond pollination in the early 1990s, he warned of its inherent contradictions. "Almond pollination has been a major source of beekeeper income in recent years and is the only reason many beekeepers are still in business," he wrote. He followed this statement with: "Almond pollination has spread parasitic pests and reduced income for many beekeepers to the point where many have been driven out of business."

When we ask him if that same contradiction holds true in these days of colony collapse disorder, when the stress of being trucked thousands of miles is cited as a possible contributing factor in CCD, he pauses, then replies slowly: "You're probably right. When you're trucking bees they need to sleep, just as humans do, and the bumping around in the truck for two to three days keeps them awake and this lowers their resistance to pests and disease."

*

Dave Hackenberg's bees have been on the road for four days. They've made the 2,600-mile (4,200km) journey west from Dade City in Florida to Lost Hills, California, 40 miles (65km) north-west of Bakersfield.

This Pennsylvanian beekeeper has joined forces with fellow east coasters Bob Harvey, from New Jersey, and Dave Mendes, from Massachusetts, who overwinter their bees in Florida. Together they are transporting 10,000 beehives on 20 semi trucks. Only 1,000 beehives belong to Hackenberg. CCD claimed two-thirds of his bees in 2006, and half again in 2008.

He says that the bees do get to sleep at night when the drivers pull in and take a nap, but when the first convoy of four truckloads arrive on Friday at 5pm the bees are so hot and bothered that each load has to be sprayed down with water at the Truck Wash. The hives are tethered with strong harnesses and covered with black netting that prevents the bees escaping and protects them from being drenched.

Hackenberg, 59, is wearing cowboy boots, checked shirt and blue jeans. He is far too talkative to be the hero of a Western. But his twangy vowels, no-nonsense language, and independent spirit are straight out of the Wild West. He even has a hard hat in the shape of a Stetson with netting attached that he wears when unloading the bees. "It's useful in case any hives fall on your head, which is not unknown," he booms.

Off the highway in a never-ending orchard, the convoy of trucks follow Hackenberg and his team, who are driving pick-ups. They are joined by a handful of fork lift trucks, slowly edging past row after row of almond trees with ground-level sprinklers spewing out water. They are guided to their destination by a strange combination: a GPS satellite navigation system, which sits atop the dashboard and recognises most of the narrow roads that run between the orchards, and a hand-drawn diagram showing the beekeeper at exactly which row of trees to place the hives, and precise instructions about how many to put there, and in what groupings.

We stop shortly past a row marked 494. The beekeepers throw on their protective suits and veils and get to work unbuckling the straps and pulling back the netting from the first truck. We expect an angry black cloud to form, but as the light fades over the orchards and the temperature drops most of the bees stay warm in their hive.

Harvey jumps on top of a fork lift and starts transferring hives 12 at a time to a pickup emblazoned with the words Harvey's Honey and the logo of a large bee. Back and forth he goes until the pickup is full, then he sits the fork lift on a small trailer hitched to the back. He drives off around the orchard, using the fork lift to place hives in groups of 20 as instructed. To avoid any confusion, three trees at regular intervals along each row carry a label stating BEE DROP in big black letters.

To unload and distribute the 2,000 or so hives across 1,000 acres of land will take a good few hours. And it's only the beginning of a gruelling week's work. Over the next few days another 16 trucks loaded with beehives will arrive in near or complete darkness. When we meet Hackenberg again in the early hours of Sunday morning he has had only a couple of hours sleep between drops.

When I ask him if this isn't too stressful for the bees, he admits: "It's got to be stressful." But he adds: "Well, everything's stressful. Getting out of bed to meet this guy at 4.30 this morning was stressful. Whether it's more stressful for the bees or for me, I'm not sure."

But it's not just the Californian trip that the bees have to endure. They can easily cover 11,000 miles (18,000km) each year, going up to the apple orchards in Washington State, then over to the north-east for cranberries and pumpkins, before finishing with blueberries in Maine in May.

While Hackenberg admits that his bees' workload "may have got more intensive" in the past few years, he does not believe it is killing them. "I've been doing this 40-odd years. I've seen accidents, small hive beetle, varroa – you name it, I've seen it. We've done all the same things, but the rules have changed. Something's messing up."

Bret Adee of Adee Honey Farms agrees that trucking is not the problem. "We've been trucking bees for 50 years, and in that time conditions have improved for the bees," he says. "The roads are smoother, the trucks are better. It takes less time to truck them around. We pay a premium for special [bee-friendly] haulage companies. We work out the route for the driver to take so they can stop at a suitable place to sleep. You can't just pull over anywhere with hundreds of beehives. We know exactly where they should be at all times."

In a typical year Adee's bees go from Californian almonds, north to pollinate apples, then down to Texas and Mississippi, where 80% of the colonies are given a new queen bee, before heading all the way to Maine for blueberries. They spend the rest of the summer on the Nebraska and South Dakota prairies, collecting nectar for honey. Then it's back to California in late October for the winter.

Ever since Mormon farmer Nephi Ephraim Miller loaded his beehives onto a railroad car in Utah in the winter of 1907, destined for the warmer climes of California, beekeepers have been on the road. Miller is credited with being the first migratory beekeeper in the US, and a century later his great-grandson, John Miller, who is based in North Dakota, keeps the family tradition alive. He has brought 11,600 hives from Idaho to the northern end of Central Valley.

Miller lost around 2,000 colonies in 2008 to the classic symptoms of CCD. He sees modern agricultural practices as the bees' worst enemy. "There are these new breed of systemic pesticides and GM corn and alfalfa that we don't know too much about," he says.

Almonds, like most crops, suffer from a number of exotic-sounding pests such as the navel orange worm, whose larvae will consume new nuts, and the peach twig bore, which tunnels into new shoots and nuts. Then there are spider mites, which cover the trees in webbing and suck the leaves dry. All these bugs are controlled by pesticides. Most farmers try to rotate the pesticides they use because insects have become very efficient at building up resistance.

Although the new pesticides Miller refers to are not used on almond trees, according to the Almond Board of California, little research has been done on the impact that rotating different pesticides could have on honeybees.

Farmers are instructed not to spray pesticides when the trees are in bloom. They do, however, use a fungicide during bloom to protect the almond tree flowers from moisture. "Careful timing of fungicide sprays will minimise bee losses," Traynor advises in his pollination handbook. Although he says fungicides are relatively non-toxic to bees, he warns that "aerial applications can physically knock down and disorient bees, and wetting bees' wings with surfactants reduces bee activity".

One fungicide with the brand name Captan is, however, described as "highly toxic to honeybee larvae". Traynor says it can reduce pollen germination and bee larvae development if larvae are fed contaminated pollen. Fifteen years ago, he predicted that the substance might soon be banned. Yet it is still on the shelves, and, according to Traynor, is used occasionally when petals are falling off trees, or after they have fallen, when it presents less danger to bees.

Traynor, who has a master's degree in soil science, doubts that the pesticides or fungicides used in California's almond orchards, or the chemical fertilisers dug into the soil, are toxic enough to poison the bees, even if they get washed into waterways from which the bees drink. Yet he, along with most beekeepers, recognises that the poor nutrition offered by monoculture is an issue for honeybees' health. The almond orchards are a barren wasteland for any bee before the bloom erupts. Even when the blossom is at its height, there is nothing else for the bees to eat. It's the same story across the pollination trail.

Adee has suggested to the almond farmers he works with that they start to plant legumes between the trees to give the bees a more varied menu. Yet farmers seem more interested in developing products that will make the bees gorge themselves on extra portions of almond blossom.

Frank Eischen is a scientist based at the federal bee laboratory in Tescon, Texas. He is in California for a few weeks conducting trials for a product called SuperBoost. The tests

are funded by Project Apis Mellifera (Pam), an initiative –
mainly paid for by almond farmers, under the auspices of
the Almond Board of California – whose aim is to "fund
and direct research improving the health and vitality of
honeybee colonies".

SuperBoost is a synthetic brood pheromone that is
claimed to increase bees' foraging activities, which means
more pollination takes place. It works essentially by tricking
the bees into thinking they need to collect more food,
telling them that there is more brood in the hive than is
really the case. At the end of each day, Eischen and his assistant, Henry Graham, pull out a drawer beneath 200-odd
hives and empty golden grains of pollen into a plastic bag
labelled with the number of the hive and the date. After
three weeks they will be able to see which colonies have
collected the most pollen. Some hives are eight-framers,
others four; some have a syrup drink in the hive; some are
broodless, others have additional natural brood to judge
how the product compares with the real thing.

It is arguable that deceiving bees into working harder
could compound the stress they already face. Another consequence of the fake pheromone could be that nurse bees,
whose job it is to care for the young in the hive, are encouraged to abandon their duties in favour of foraging outside.

When we ask Eischen why he is carrying out these product tests when much more research is still needed to find the

cause of CCD, he replies in his measured tone that he is an "unbiased advisor" who is funded to do it.

Since 1974, the Almond Board of California has paid for more than $1.4 million (£1 million) of bee research, predominantly to develop nutritional supplements that will boost bees' productivity during the almond pollination. In 2008 alone it gave $200,000 (£144,000) for bee research.

Alongside the field trials is a tent with eight observation hives, each holding eight or four frames of bees. The idea here is to test whether foragers from big colonies have time to seek out the few alternative food sources available outside the orchards (from, say, the almond ranchers' gardens), while smaller colonies must stick to the nearest meal in the branches of the almond trees.

With the price of strong beehives accounting for a fifth of almond farmers' costs, and showing no signs of abating, it is not difficult to see why growers are keen to find a product and a smaller colony size that does the same job more cheaply. But none will admit this is cost-cutting thinly disguised as improving honeybees' health.

Deceiving bees is at the heart of the biggest pollination event on the planet. California's almond trees bloom so early in the spring that they demand colonies of bees to be far stronger than they would normally be so soon after winter. Since beekeepers' livelihoods rest on their bees' performance for three weeks in February, they dupe them into thinking it's already summer by feeding them an array of protein and energy supplements. The more food that comes into the hive – usually brought back by foragers in summer – the more eggs the queen lays, and more eggs create more workers to go out to forage and pollinate.

Says Traynor: "We're interfering with their natural cycle because we want strong colonies for almond pollination. We're stimulating hives in August, September and October and making the queens do a lot more laying. As a result the queens are suffering burnout. It used to be that a beekeeper could pretty much leave his bees alone during winter. That's no longer the case."

Manufactured protein supplements often come in the form of patties, which are a mixture of sucrose, brewer's yeast and pollen, often imported from China. Some beekeepers also feed their bees egg whites, which are a similar consistency to royal jelly. An energy booster drink is made of corn syrup pumped into a plastic feeder in each hive. It doesn't come cheap. Ron Spears, one of the largest bee operators in the States, tell us that he spends up to $34,000 (£24,000) a year just on liquid sugar for his 18,000 colonies.

The quality of the manufactured food supplements is variable, and natural chemicals in corn syrup have been found to become very toxic to bees if heated above a certain temperature, as could happen during storage or transportation.

Another of Pam's projects is investigating the problems associated with high-fructose corn syrup.

Eischen has studied the best time to feed bees. He found that bees in California that were fed all the way through winter were in great shape for almond pollination. But what about winter feeding's long-term impact on the queen, who is denied a rest from laying? "There may be longevity issues," Eischen admits.

Dan Cummings, chair of the Almond Board of California's bee task force, believes that what is good for the almond growers is good for the bees. He told delegates at the 2008 national beekeepers conference in Sacramento that the solution for the almond industry (problem: high hive rental prices) and the bee industry (problem: dying bees) is the same – "hives populated by strong bees to provide a pollination service to the almond industry". Yet the Almond Board of California, more than anything else, is responsible for creating so many demands on honeybees.

Established in the 1950s by a federal marketing order, the ABC receives a three-cent mandatory levy per pound of almond crop, which means that last year it enjoyed a $40 million (£29 million) budget, two thirds of which went on marketing almonds at home and abroad. Its phenomenal success at tapping new markets and promoting almonds' nutritional and health benefits has fuelled a staggering growth in demand of 6.8% every year for 25 years. And there is no let-up.

Although China is thought to be considering going into almond production, the Almond Board forecasts that by 2011 Central Valley's almond orchards will cover 755,000 acres (305,000 hectares) – an extra 150,000 acres (60,000 hectares) in just three years – and will require 1.6 million honeybee colonies for pollination.

The figures don't surprise Hackenberg. "They are planting almonds out here at the most ungodly rate you ever seen in your life," he says. "South Valley Farms down here are going to plant 10,000 more acres; Paramount [the largest almond farmer with 40,000 acres] has still got thousands to plant."

Cummings assures us that the Almond Board's projected growth is achievable even though colony collapse disorder wiped out a third of honeybees across the States in 2006-7. "It's a matter of supply and demand. If there's not enough supply, the price of renting bees goes up and more bee-keepers will bring theirs to California," he explains.

Yet he wasn't so sure last year, when he was quoted in an article as saying "about 70% of the transportable colonies in the US are coming into the almonds right now, and we're going to have about a 25% increase in bearing acreage in the next four to five years. So do the math. We're going to have a problem."

Cummings, a Harvard-educated almond and walnut farmer with a share in a beekeeping operation, believes colony collapse disorder is "highly correlated with stress". "It's just like a person who is strong and healthy being able to fight off a flu bug, but it becomes a serious condition for someone who is weak," he says. Yet he rejects any notion that the bees have buckled under the combined pressures inflicted by almond pollination: long-distance travel, poor nutrition and interference with the bees' natural cycle.

"Beekeeping is a migratory business," he says "Beekeepers are constantly moving bees to follow nectar flows. Now we also do it for pollination, and as almonds are the first main source of nectar and pollen in the year the bees come out much stronger."

On monocultures, he points out that bees naturally feed on just one crop at a time as it blooms. As for the bees' cycle, he describes it as "fairly dynamic", and "not a precise cycle" when food is available. "Bees in the US live in a wide range of climates, and for some – such as those in Hawaii, where many of our queens are bred – there is no winter."

California's almond orchards in spring have been compared to a major crime scene, or a big brothel where pests and disease are easily spread. Cummings agreed with the analogy last year when he told a reporter: "You've got bees coming from all over the US with all of their respective regional problems and they're all being put in close proximity in California's almond orchards in the spring. It just stands to reason that any malady anywhere in the US is going to come to California and affect the hive."

Now all he will say to us on the matter is: "I don't know what difference it makes to CCD."

Since the US lifted its ban on importing bees from Australia in 2004, diseases and viruses carried by Antipodean bees have been added to the melting pot. Thousands of packages have been flown in from Sydney to make up any shortfalls for almond pollination. They land at San Francisco airport and are dumped into hives destined for Central Valley orchards.

Some beekeepers have gone to considerable lengths to protect their bees from mixing with outsiders. "I try to isolate my hives from the east coasters," says the California-based Spears. "I have 3,000 of my own hives next to each other in the San Joaquin Valley [which covers the southern end of Central Valley]."

Adee adds: "We used to have our bees in isolated pockets, but as acreage grows pockets disappear and now a neighbour may have Australian stock."

Australian honeybees are among those carrying Israeli acute paralysis virus, which has been identified as a marker for CCD, so beekeepers are understandably nervous.

Scientists are not yet sure how CCD could be transmitted between bees from different colonies. One theory is that if CCD is caused by a virus, the virus may be transmitted through the bees' saliva on to pieces of a flower's pollen, some of which will be collected by other bees and taken back to their hives.

Jerry Bromenshenk's company Bee Alert Technology sells hive-tracking devices. In 2007 it sponsored a survey that showed CCD had reached 35 states. Bromenshenk, who is also a research professor at the University of Montana and one of the original members of the CCD working team, says he has seen CCD spread across a bee yard in a matter of weeks, taking out 50% to 80% of the colonies. "In large holding yards, we've seen it start at one end and roll through to the other end like a wave."

Holding yards are usually barren valleys or fields, away from potentially harmful crop spraying, where beekeepers place thousands of colonies to get them ready to pollinate almonds, like the remote desert area where Adee had his bees. Hives – which may be trucked in from a number of states – are generally close together and, as in crowded quarters everywhere, disease can easily spread. "We saw CCD move through holding yards from one end to the other as early as December 2006," Bromenshenk says. "We've been all over the US looking at CCD and have often seen it move through holding yards and apiaries."

The real question, however, is whether the pollination industry is making the bees too weak to fight a virus that is being passed around.

Marla Spivak, entomologist at the University of Minnesota, says: "We're placing so many demands on bees we're forgetting that they're a living organism and that they have a seasonal life cycle and they're going to have down times. We're wanting them to function as a machine. We want them to be strong and healthy all the time, and we're transporting them like they were a machine ... and expecting them to get off the truck and be fine."

What we are witnessing in California is no less than the industrialisation of pollination. Bees here are continually referred to as livestock – something we'd not heard before. It's as if they have become battery bees. As Tom Theobald, a Colorado-based beekeeper, says: "We have pushed the bees into an industrial age."

Eischen says beekeepers are now likened to "a bunch of mechanised gypsies". He points out that livestock can get an illness called shipping fever when they are transported. "It's fairly common that they can get sick. We don't know if bees get sick but we should look into it. It could be important." Miller says modern agriculture is pushed harder than it ever used to be and is being kept aloft by "ever more fragile wings".

For 2008 at least, the wings managed to continue beating. After getting off to a slow start, the trees finally burst into flower by the last week of February and Central Valley was covered in a snow-like haze. Near-perfect weather made for ideal pollination conditions, the warm, sunny days encouraging even the weakest colony to forage for long hours.

Hackenberg had been predicting that the growers weren't going to have enough bees when the bloom arrived. "We're filling in bees across the road tomorrow night for guys sitting in the parking lot who are going to be short," he told us during one of his drops. "The brokers and growers don't get told they are short until ... Guess what! I came up 400 short, 200 short, 600 short. After a while it starts to add up into thousands."

Yet his fears didn't materialise, even though many of the beekeepers we spoke to on their way to Central Valley were bringing only half as many bees as they'd expected because of colony collapse disorder. Only Spears seemed to have escaped the plague. His secret, he claims, was moving his bees from drought-ridden California up to North Dakota where they could forage for food and water last summer. Bret Adee was lucky enough to be able to call on friends and colleagues this year to plug his huge losses. Others, such as Miller, covered themselves by bringing 15% more hives to California than their contracts demanded. Traynor reported that only 100 of the 35,000 hives he brokered ended up with no bees. By spotting the problem early, he managed to find replacements.

Cummings says growers don't really need as many as two hives per acre to pollinate their almonds: it's an insurance policy against what they call "dead-outs" – empty hives. But what insurance does the environment have against the billion-dollar almond industry if its practices are contributing to the demise of our most effective pollinator?

As the sun rises over the orchards after another nocturnal rendezvous, Hackenberg says it's only the money that brings him and his bees to California each year. "I'd rather be back in Florida with my bees. They'd be feeding on the maple and willow. It's paradise down there. Why would anyone come to this God-forsaken place? But something's got to pay the bills. I'm here for a $150,000 cheque."

CHAPTER 10

A WORLD WITHOUT BEES

The mountains of southern Sichuan in China are covered in pear trees. Every April, they are home to a rare sight: thousands of people holding bamboo sticks with chicken feathers attached to the end, clambering among the blossom-laden branches. Closer inspection reveals that children, parents and even grandparents are all pollinating the trees by hand. It is a ritual they have been following for more than 20 years, ever since pesticides killed their honeybees.

Local farmer Cao Xing Yuan, interviewed for the US documentary Silence of the Bees, said that when the bees disappeared he wrote to Beijing asking what should be done. He was told that humans would have to take their place.

That's a tough job. The farmers must first collect pollen from the trees by scrubbing it off the anthers, the male part of the flowers, into a bowl. They let it dry for two days, then the whole family comes out with their homemade feather dusters, which are dipped in the pollen and applied to the flowers' stigmas, or female parts. It is a slow, laborious process and much less efficient than a colony of honeybees, which could visit three million flowers in a day. But the hand pollination seems to work. In August, the trees are heavy with fruit and each family harvests around five tonnes (11,000lb) of pears.

If honeybees were to disappear off the face of the planet, would we all be reaching for our feather dusters? It's unlikely. In the US, to employ even low-wage labourers to hand-pollinate the 3.5 million acres (1.4 million hectares) of crops normally fertilised by honeybees would cost an estimated \$90 billion (£64 billion) a year. Even in China, there are fears that as more and more young people leave the countryside for the towns and cities in the next decades, there will be no one left to pollinate the pears.

Rachel Carson imagined fruitless autumns in her book Silent Spring, about the environmental destruction wreaked by pesticides: "The apple trees were coming into bloom but no bees droned among the blossoms, so there was no pollination and there would be no fruit." Today it is more than pesticides that threaten honeybees' survival. But could buzzing on a summer's day really become a distant memory?

The National Research Council's committee on the status of pollinators in North America warned in 2006, even before colony collapse disorder had claimed a third of all US honeybees, that the much-loved managed insect could be extinct in less than 30 years. "The US commercial honeybee population was stable from 1996 to 2004, but if it were to continue to decline at the rates exhibited from 1947 to 1972 and from 1989 to 1996, it would vanish by 2035," said the committee's report. But is there evidence that commercial beekeeping is on its way out?

If there was a golden age of beekeeping in the States, it was just after the second world war, when 5.9 million colonies were kept. By 1988, the number had almost halved to 3.4 million. Today there are 2.44 million colonies – less than half that of 1947 – predominantly kept by the 1,000 commercial operators.

The downward trajectory is a result of the numerous assaults on honeybees, ranging from mites and fungal infections to the Africanised bees that took up residence in the States in 1990. It is predicted that beekeepers in the southern half of the US may have to abandon their hives as the aggressive Africanised hybrid takes over, because they won't be able to afford the insurance to cover the lawsuits over bee stings.

But it's not just in the US that beekeeping has taken a nosedive. In England and Wales the peak of ownership was also 60-odd years ago. From August 1947 to January 1948, there were more than 360,000 hives, according to figures collected by the then Ministry of Food, which provided some 76,000 beekeepers with allowances of sugar to feed their colonies. Will Messenger, a beekeeper who has set up a beekeeping history group, points out that not everyone claiming to be a beekeeper at the time was genuine. But he says the ministry figures should be reliable as it sent out its inspectors to check all claims of bee ownership. "There are references to prosecutions for false claims of ownership of colonies of bees," says Messenger. "A delightful picture is created of desperate housewives eager to make jam trying to fool the ministry official into believing that they had colonies of bees that required the sugar ration. Hives were opened and found to contain dead bees, or perhaps a handful thrown in just before the inspector arrived."

Now there are only around 270,000 hives across the whole of Britain, and just 44,000 beekeepers. With increased winter losses reported, it seems unlikely that there will be a recovery any time soon. A 2008 survey by the department for the environment, food and rural affairs found that one in four of the colonies its bee inspectors had been called out to see didn't make it through. The UK government denies, however, that CCD has reached British shores and attributes the heavy loses to varroa mites. Even so, farming minister Lord Rooker has predicted the demise of the honeybee in Britain within a decade. In November 2007, he told parliament: "We do not deny that bee health is at risk. Frankly, if nothing is done about it, the honeybee population could be wiped out in 10 years."

Statistics tell a similar story of losses across Europe, where a variety of assailants, from varroa to pesticides and fungi, are being held responsible. Germany, for example, had four million hives a century ago. Now it is down to 800,000 – an 80% drop. "If that continues there will eventually be no bees," says Professor Jürgen Tautz, a bee expert from Würzburg University.

In the States, results from an Apiary Inspectors of America (AIA) survey of beekeepers' colony losses across 29 states revealed a 36% mortality rate over the 2007-8 winter. More than a third of operations said that at least some of their colonies died with the "absence of dead bees" symptom that is the hallmark of CCD, although the survey concluded that at least 71% of all colony deaths were *not* caused by CCD.

Dennis vanEngelsdorp, the Pennsylvania state apiarist and AIA president, described the losses as "alarming". "Beekeepers have lost up to one million colonies this winter. More west coast operations have been hit this year," he said. "It's particularly alarming that so many bee deaths are down to known causes such as varroa."

Ask American beekeepers how much longer they will be in business, and you get answers like: "As long as the bank continues to bail me out" and "My children have all gone to college. They don't want to run the business when I retire in a few years time. So it will probably go with me." As more and more hives have to be restocked each year, bankruptcy stares beekeepers in the face. And with the average age of the commercial guys (it's always guys) pushing 60 and few youngsters coming through the ranks, the future looks bleak. The hobbyists may continue to manage the odd hive, but even they may be discouraged if the killing spree continues, vanEngelsdorp warns. The AIA survey found no difference in the proportion of bee losses in 2008 between large-scale migratory beekeepers, medium-sized operators and small outfits.

If all beekeepers deserted their hives tomorrow, would the inhabitants die? Judging by what we know has happened to feral honeybee colonies in the US since the arrival of varroa, the answer is yes. In fact, almost all unmanaged western honeybee colonies around the world – outside varroa-free Australia – would be wiped out by the parasitic mite. The few that did manage to fight off their foe would be too small in number to pollinate the global food chain.

It is hard for people to grasp the full horror that would ensue if honeybees did vanish. May Berenbaum, chair of the committee on the status of pollinators in North America, notes that their threatened demise is unlikely to inspire a rash of Hollywood disaster films. "Tidal waves, floods, fires and explosions still remain inherently more cinematic than just about anything involving flowers, birds, bees and butterflies," she says.

The nearest we got to a filmic depiction of disappearing honeybees was a colourless Central Park in Jerry Seinfeld's animated comedy Bee Movie – hardly the stuff of nightmares. But it's not just pretty flowers that would wither and die.

Most people's initial response to the idea of a world without bees is: "That's a shame – I'll have no honey to spread on my toast", or, "Good – one less bug that can sting me." Yet without the insect that pollinates many of the plants we rely on for food, beekeepers warn of an economic and ecological disaster.

Hardly a week goes by without a new doom-filled report warning us how man-made climate change will destroy the world as we know it. They paint terrifying pictures of melting ice caps and rising sea levels submerging whole nations; decades of droughts and water shortages; starvation and wars across the globe as people fight over ever-scarcer resources. But doomsday scenarios should not be reserved for global warming.

Some 12% of the Earth's land mass is cultivated for the growing of crops. Add grazing land, and more than a third of terra firma is dedicated to human food production. How much of that would stay intact if the bees disappeared? Not the yellow fields of sunflowers or rapeseed stretching to the horizon, nor the fruit-filled orchards whose offerings fill the bowls on our kitchen tables, or the clover-rich pastures and the dairy herds that graze upon them.

The US congressional research service estimates the annual value of the 11 crops most dependent on honeybee pollination at $11.86 billion (£8.5 billion). Alfalfa, which is turned into cattle feed, tops the table at $4.7 billion (£3.4 billion), followed by apples, almonds, cotton, citrus, soya beans, onions, broccoli, carrots, sunflowers and melons. Other crops almost entirely dependent on honey-bees include blueberries, cherries and pumpkins. How long before these crops all shrivel and the landscape turns from fruitful fields to barren wasteland?

When the early European settlers arrived in America, there were no meadows for grazing livestock, nor ploughed fields to grow crops, and no honeybees to pollinate them. In his book Changes in the Land, William Cronon charts the transformation of Massachusetts' forests from a "bushy wild-woody wilderness" into a "second England for fertile-ness". European farmers and their descendants cleared three-quarters of New England's forest for farmland. Tall, thick-trunked trees were felled to build strong ships and timber-beamed homes and provide plentiful supplies of fire-wood. In their place were planted many bee-pollinated crops and pastures. Wolves, beavers, deer and turkeys were soon replaced by sheep, cows and pigs. An experiment to see what would happen to the land today if it were allowed to revert to its natural state has witnessed ash trees, white pine and sugar maples reborn on former farmland.

So would the landscape in a world without bees revert to forest? Probably not. Wheat-washed plains, corn-covered prairies and flooded paddy fields would still create a patch-work of agriculture across the surface of the globe, sustained by wind pollination. But the pattern would lose its cotton plantations, vegetable beds, orchards and acres of alfalfa grown for cattle forage.

How would our weekly food shop change? With honey-bees responsible for so much of what we eat, the list you scribble before the trip to the supermarket would get shorter and become less palatable. Off comes the honey, followed by fruit, save for bananas and pineapples, and most vegetables, along with protein-rich beans, meat and dairy products.

It sounds odd when you say no bees equals no steak and no bacon. The connection is not obvious, but without bees to pollinate crops grown for cattle and pig feed that could happen. And it won't just be joints of meat that go: cheese, milk and even ice cream could disappear, or become prohib-itively expensive.

Honeybees also dramatically increase yields of coffee, so without them the shelves of rich-smelling beans would be sparsely filled. And where dozens of types of cooking oils used to stand, only a couple – walnut and olive – would remain. The fish counter might still be stocked – but with fewer sources of protein available the seas will probably have been plundered.

That leaves bread. But what do you spread on it? Rice and pasta are plentiful, but where are the ingredients to make a tasty sauce? And let's hope you can get used to pizza with no cheese topping, taco shells without the refried beans, and noodles but no bean sprouts.

Breakfast will consist of a dry piece of toast, a bowl of porridge made with water, and an egg. No fruit juice to wash it down. You couldn't even substitute soya milk for cows' as the bean from which it is extracted relies on bee pollination.

Think about what you have just had for lunch: a sandwich, a bowl of soup or a salad, perhaps. How much of that would have been possible without vegetables or a slice of cheese? And tonight's meal: how would you prepare that with just cereals and grains?

Tim Lovett, president of the British Beekeepers Association, in a plea to secure £8 million ($11 million) for bee health research over the next five years from the UK government, asked ministers: "Does the government want the nation to go without honey on their toast, not have home-grown strawberries to go with cream, and even put their own crusade for the public to eat five portions of fresh fruit and vegetables at risk?" But it's not just our diets that would change beyond all recognition if bees were to vanish. We would also have to give up the clothes on our backs, from T-shirts and jeans to chinos and denim skirts. The cotton plant that clothes us has far higher yields when pollinated by honeybees.

Medicines also rely on flowering plants pollinated by bees. Digitalin, a drug that treats irregularities in heart rhythm, comes directly from foxglove flowers; the decongestant ephedrin is from the shrub ephedra; and reserpine, which lowers blood pressure, is made from serpent-root. Chemicals are also extracted from plants and used as building-blocks to create new compounds. Etoposide and teniposide, which treat skin cancers and warts, for example, are manufactured from epipodophyllotoxin, a chemical found in May apple.

Plants also feature heavily in the lexicon of alternative medicine. The herbal industry promotes thousands of "natural" remedies for common ailments, ranging from evening primrose oil, made from the crushed seeds of the plant, for combating pre-menstrual tension, to the flowers of St John's wort that are used in teas and tablets to relieve anxiety and mild depression.

Beeswax has over 120 industrial uses in drugs, polishes, lubricants and skin care products. In 2005, the European Union consumed 8,000 tonnes (18 million pounds) of the substance, 60% of which went into pharmaceuticals and cosmetics, where it acts as an emollient, emulsifier and stiff-ening agent for oils and fats. Some of its more obscure uses include coating the strings on archery bows, and water-proofing whips.

Yet if bees vanished tomorrow and with them their wax and bee-dependent plants, wouldn't we just find alternative ways to feed, clothe and cure ourselves? Paraffin has already superseded beeswax in most applications because it is cheaper and easier to produce, while nylon, rayon and poly-ester are among the synthetic fibres we could wear.

Imagining, however, that science will somehow come to our rescue shows a spectacular failure to comprehend the scale of the crisis we could face if nature's master pollinator spiralled into extinction, from the unravelling of the world economy to the collapse of the terrestrial environment.

A former US agriculture secretary, Mike Johanns, warned in June 2007 that CCD could cost the US economy as much as $75 billion (£54 billion) in lost output and jobs. If the US lost all of its 2.4 million honeybee colonies, but they remained intact in other countries, its food manufactur-ers and supermarkets have said they would weather the storm by importing more foodstuffs from overseas. It is esti-mated that by 2012 a quarter of the vegetables consumed in the US will already come from China and in 50 years the US will be a net importer of food.

If honeybees were wiped out in Argentina, the world's largest exporter of honey would lose an industry worth $134 million (£96 million) a year. In Europe, the annual cost would be around 400 million euros ($615 million/ £440 million) excluding pollination services, and China, the biggest exporter of royal jelly, beeswax and other bee products, would also lose a lucrative export. Big honey and beeswax producers such as Mexico, Turkey and Ethiopia would also be badly hit.

The UN's Food and Agricultural Organisation puts a figure of $200 billion (£144 billion) on the annual value of pollination services worldwide (although this is not just for honeybees). If honeybees vanished off the face of the earth, neither the US nor any other country would be able to solve food shortages by importing more fruits, vegetables or cattle feed.

Garth Cambray, a South African manufacturer of honey-based products, describes the economic meltdown that would ensue. In China, for example, pork is a major component of food bills, so a fall anywhere in the world in the production of the seeds that are used in pig feed would not only lead to a shortage of pork, and push up prices for the meat, but also increase inflation. This would force the Chinese government to increase interest rates to try to contain it. Because the Chinese and US economies are now so intertwined, inflation in China would push up

interest rates in the States and the whole global credit economy, which is already in a parlous state, would suffer a
further blow.

In addition to the financial turmoil, what about the environmental catastrophe? In his book The Creation, the
celebrated biologist EO Wilson outlines the devastating
impact on the environment if all insects were to vanish.
Firstly, a majority of flowering plants fail to reproduce. Most
herbaceous plants die out. The great majority of birds and
mammals, reptiles and amphibians, denied their fruits, foliage
and insect prey, follow plants into oblivion. Soil remains
unturned by insects, accelerating plant decline. Populations
of fungi and bacteria explode, feeding on the dead and
rotting vegetation. Wind-pollinated grasses and fern and
conifer species spread over much of the deforested terrain.
Humans survive, able to eat wind-pollinated grains and
marine fish, but there is widespread starvation and wars for
control of the dwindling resources, while human populations
plunge to a small fraction of their former level. "The tumul-
tuous decline to dark-age barbarism would be unprecedented
in human history," Wilson says. "The survivors would offer
prayers for the return of weeds and bugs."

Although Wilson's dark vision follows the loss of all
insects, without nature's premier pollinator the picture looks
similarly frightening. Just a basic understanding of food
chains explains why removing honeybees from the equation

would break the link that begins with flowering plants. Without the flowers being pollinated, there would be far fewer seeds, roots, leaves, flowers or fruits for birds and small mammals to eat and they would die. As a result, their predators – the omnivores or carnivores that continue the chain – would starve.

Honeybees are raised for large-scale pollination because they are numerous and easy to manage and transport, but we are told that there are more than 20,000 or so other bee species around the world, 17,000 of which have been formally identified. These include sweat bees, squash bees, carpenter bees, andrenid bees, mason bees, leafcutter bees, cactus-loving bees and, of course, bumblebees. So if honeybees suffer continued decline, could these solitary bees step into their shoes and prevent our downfall?

The federal bee biology and systemics laboratory in Logan, Utah, is the only scientific institution in the US devoted to providing knowledge on how to manage non-honey bees. It convened the first international workshop on their potential role as crop pollinators in 1992, and is conducting tests to see which among the 4,500 wild solitary bee species in North America could become managed pollinators.

The lab is working with unsocial creatures that live alone in holes in the ground, wood, soft mortar, or hollows of reeds and cane. They have neither hives nor honeycomb. The alfalfa leafcutter bee, as its name suggests, pollinates the alfalfa plant while cutting its leaves into pieces to make nests for its young. The metallic emerald *Osmia aglaia* bee makes self-pollinating blackberries and raspberries much plumper and more plentiful. According to studies conducted in Oregon and Utah, when this bee is introduced into an area it visits nearly as many flowers as honeybees and makes the berries 30% heavier. Nesting boards are placed near the fruits to encourage the wild bees to stay in the fields, but there is no guarantee that they will return to these manmade homes, so researchers are investigating the way in which bees respond to different scents, in the hope that they will be able to lure them home.

The gourd bee outperforms the honeybee on squashes, pumpkins and gourds in the deserts of Arizona and northern Mexico. By the time honeybees take flight and arrive in the squash blossoms in the morning, the earlier-flying gourd bees have already got the metaphorical worm, or nectar in this case. Studies show it would take an average 3.3 visits per honeybee but as few as 1.3 visits per gourd bee and 1.1 visits per carpenter bee to fully pollinate each female flower.

The wild bee species with the most commercial potential is the blue orchard bee, *Osmia lignaria*. It can fly in cooler weather than honeybees and has proved an excellent pollinator for a long list of fruits including apples, pears, cherries, plums, peaches, nectarines, strawberries and almonds. It visits more flowers per minute than a honeybee and more trees in a day of foraging, so only 250 females per acre are needed to pollinate apples, and 300 per acre for almonds, compared with thousands of honeybees.

The Utah lab has bred shelter-living blue orchard bees to begin commercial almond pollination. It claims that in 10 years' time, the iridescent blue bee, which is equipped with a dense brush of hair on the underside of its abdomen, may be able to pollinate half the almond crop. However, it still has to figure out how to keep the bees in the shelter as the queen instinctively seeks out a new nesting site each year to lay her eggs. These solitary bees also need to be manipulated to emerge from winter two months earlier than nature intended in time for the February blossoms. So for the foreseeable future it is likely to be much easier to get two hives of honeybees, each containing 30,000 bees, to pollinate an acre of almond orchard than to somehow press into service a few hundred solitary, wild bees.

Bumblebees are no slouches when it comes to pollination. This large, spherical, fluffy bee, of which there are 239 species worldwide, is adept at pollinating plants that will only release pollen when parts of their flowers receive intense vibration. Tomatoes, potatoes and artichokes all rely on the bumblebee's trembling action to reproduce. The bumblebee grips the flowers of the plant and shivers its flight muscles without opening its wings. In the US, this service is valued at $3 billion (£2.2 billion) a year. But wild bees are not a panacea for a world without honeybees, because their fragile existence is also threatened.

According to Stephen Buchmann and Gary Nabhan, in their book The Forgotten Pollinators, America's lowbush blueberries were historically pollinated by at least 190 kinds of native bees, depending on the state in which they were grown. Now they are reliant on honeybees being trucked in, because overzealous weed control by blueberry managers has resulted in the loss of habitat and forage for the native pollinators. It is not an isolated case.

Long-term population trends for bumblebees are demonstrably downward. Although wild bumblebees can pick up pathogens from commercially bred bumblebees imported from Europe to pollinate greenhouse tomatoes, the overwhelming reason for their decline is habitat loss. Changes in agricultural practices have led to huge areas of land being planted with a single crop. The removal of hedgerows and field margins in this monocultural landscape has robbed native pollinators of somewhere to nest, while the abandonment of crop rotation, in favour of fertilisers, and the elimination of weeds in these huge fields and pastures have both contributed to a dearth of food.

The loss of flower-rich grasslands seems to underlie the

decline of at least three previously common bumblebee species in Britain. Grazing can disrupt ground-nesting bees, affect availability of water and nectar, and reduce the diversity and abundance of floral resources. Excessive mowing of embankments, road sides and public areas also leads to loss of flowers and nesting sites.

Half of the bumblebee species in the UK are either already extinct or could face extinction in the next few decades, according to a report on the status of native pollinators in North America. "The apparent loss of two species in the US in the past few years suggests that North American bumblebees are similarly imperilled as a result of the combined effects of numerous anthropogenic [man-made] factors, including habitat loss, degradation, conversion, pesticide use and pollution," it concluded. The report called on the US government to offer farmers and ranchers financial incentives to manage their land in a more pollinator-friendly way.

The Xerces Society runs the California pollinator conservation project. Since 2003, it has been providing farms in Yolo County with a mixture of plants that flower throughout the year, and nest blocks for bees, as well as keeping large areas of untilled soil where native bees can live. The sites are being monitored to document how effectively they are encouraging back native bees and restoring pollination services to the environment. The federal-funded project is also training local "citizen scientists" to identify pollinators and monitor sites.

Judith Redmond, a partner at Full Belly Farm, an organic produce grower in Yolo County's Capay Valley, told the San Francisco Chronicle that the measures had made a considerable improvement. "Very clearly from our farm experience, the habitats that we've installed here have made a difference to the pollinator population," she said. But final details being hammered out in the farm bill on Capitol Hill are looking to trim conservation budgets, which will jeopardise projects like the one in Yolo County.

Until 2007, the European Union had been paying its farmers to take some of their land out of production, partly to reduce its infamous "butter mountains" and "wine lakes". This system of "set aside" was also designed to try to undo some of the damage that intensive farming had caused to ecosystems and wildlife. Farmers received subsidies for planting up to 30% of their land with pollinator-friendly ryegrass, mustard or clover. These EU subsidies allowed farms to become places for recreation and conservation as well as agriculture.

Fir Tree Farm, in St Helens in the north-west of England, for example, used to be an arable farm growing 39 hectares (97 acres) of cereals, potatoes and sugar beet. From 1989 to 2003, the owners, Alan and Edwina Abbott, planted 13 hectares (33 acres) of woodland under the

EU-funded Mersey Forest scheme. A further 24 hectares (60 acres) is now under the Farm Stewardship Scheme and provides low-density grazing for six highland cattle on meadows of wildflowers. The change in the landscape, they say, has led to the reappearance of wildflowers and birds, including barn owls and stonechats, and they have noticed a big increase in insects such as dragon and damsel flies, butterflies, moths and bees.

Other habitat restoration schemes across Europe focus on protecting a variety of natural environments such as cliffs, lowland heaths, marshes, sand dunes and levees adjacent to farmlands. Funding wildlife protection has been found to help sustain agricultural productivity.

One of the problems facing the States, however, is the huge gulf between uncultivated countryside and agricultural land. The 1964 US Wilderness Act was a historic piece of legislation that preserved 9.1 million acres (3.7 million hectares) "for the use and enjoyment of the American people in such a manner as will leave them unimpaired for future use and enjoyment". It mandated the protection of huge tracts of unspoilt land and water such as Montana's Great Bear Wilderness and the Allagash Wilderness.

These national parks are a haven for wildlife, including native pollinators such as butterflies, moths, bats, birds and wild bees. But the segregation of areas of the US into "wilderness" and agricultural land means that these creatures are too far away from the crops they could help to pollinate. Cashew nuts, squash, mangos, cardamom and cacao (chocolate) are among the crops that are pollinated primarily by wild insects. But because their habitats are too far away from farmers' fields, honeybees are trucked in to do the job instead.

If wild bees are encouraged to work alongside honeybees, studies show, honeybees themselves will pollinate five time more flowers in a day, simply because there are other bees around. Researchers Sarah Greenleaf and Claire Kremen reported their findings on sunflowers in the 2006 Proceedings of the National Academy of Sciences. Their results, they said, suggested that conserving wild habitat and altering selected farm management techniques could increase production. Expanding food production by encouraging more wild bee pollination in this way could also help to reduce our overdependency on honeybees, say Greenleaf and Kremen.

The honeybee die-off could not have come at a more inopportune time, with food production plummeting and prices soaring. In 2007, the price on the global markets of soya beans rose by 87%, corn by 31% and wheat by 130%. Escalating prices for staple foods across the globe have provoked riots in more than 30 countries.

The world is starting to wake up to the fact that its current agricultural system will be unable to feed a population that is predicted to expand from 6.7 billion to 9.2 billion by 2050. In April 2008 a report backed by 60 countries, the World Bank and most UN bodies called for a radical shake-up of world farming to avert increasing regional food shortages, escalating prices and growing environmental problems. "Rising populations and incomes will intensify food demand, especially for meat and milk, which will compete for land with crops, as will biofuels," said the authors of the 2,500-page report by the International Assessment of Agricultural Science and Technology for Development (IAASTD).

Robert Watson, director of the IAASTD and chief scientist at the UK department for environment, food and rural affairs, said: "Agriculture has a footprint on all of the big environmental issues, so we have to make sure that we don't degrade our soil, we don't degrade the water, and we don't have an adverse effect on biodiversity." He adds: "We are putting food that appears cheap on our tables but it is food that costs us dearly in terms of water, soil and the biological diversity on which all our futures depend."

Without the realisation that agriculture can help to protect biodiversity, he says, we risk undermining our ability to produce the food we need in a cost-effective way.

It is not unknown for civilisations to die of starvation. In Collapse, his best-selling account of how societies failed, Jared Diamond attributes the downfall of South America's ancient Mayan culture to its inability to grow enough food crops because of environmental damage and population growth. Deforestation, hillside erosion and the depletion of soil nutrients from overfarming led to a reduction in the amount of usable arable land at a time of a population explosion. Food shortages led to war as Mayans fought each other for diminishing resources.

Is this what the honeybees are telling us? That our industrialised farming with its monocultures, pesticides and increasingly unreasonable demands on honeybees themselves is not sustainable? With their limited resistance to poisons and pollutants, are they the canary in the coal mine warning us that if our lifestyles are killing them, we are not far behind?

EO Wilson has said that the intense media interest and speculation around the mysterious, and potentially catastrophic, bee die-off has helped to raise awareness that our very existence depends upon bugs. We are inextricably linked. He sees the honeybee as the flag-bearer for the whole of the insect race. But she is also raising the alarm for humankind.

CHAPTER 11

LAST CHANCES

It is now more than two years since CCD was discovered by Dave Hackenberg in a Florida bee yard. Billions of honeybee deaths later, what have we learned?

Scientists have still not identified the assassin. They do, however, have a pretty good idea that it is viral – like flu – perhaps with a bacterial henchman, and has a number of potential accomplices. These include pesticides, mites and poor nutrition that together attack the bees' immune system, allowing the pathogen to do its deadly work.

When we finished the first edition of this book in May 2008, excitement surrounding the discovery of Israeli acute paralysis virus (IAPV) was already on the wane. Now, a year later, IAPV is seen as just one of a number of common honeybee viruses found in colonies with high mortality rates. Although the biotech company Beeologics is trialling an antiviral vaccine in the States, attention seems to have shifted from finding a cure to eliminating some of the causes.

We argued a year ago that trying to create genetically modified superbees that could fend off parasites and disease, or products that tricked fewer bees into working harder, ignored what the bees were trying to tell us: that we need a fundamental rethink about what we expect from them, how we look after them, and what we are exposing them to.

We have seen honeybees treated more like machines than animals. Year-round pollinators and honey-making machines? This is not what nature intended. If we treat animals like automata, then we shouldn't be surprised when they break. We contended that honeybees had been pushed to the point of collapse so that the agricultural system could keep producing cheap food around the world.

Environmental changes, from urbanisation to rising temperatures and the spread of monocultural farming, we stressed, had made the honeybees' diet less nutritious, thereby weakening their defence mechanisms. And we questioned whether the laws governing the approval and use of pesticides and fungicides took honeybee health seriously enough. We called for funding for further research into the impact of sublethal doses of these chemicals on the whole honeybee colony.

When we spoke to Jeff Pettis, co-ordinator of the USDA's five-year programme to improve honeybee survival, he had raised concerns about the pesticide chlorpyrifos and two fungicides, myclobutanil and chlorothalonil, that were found at surprisingly high levels in pollen samples in tests done at Pennsylvania State university. Pettis had said he wanted to find out what the possible effects could be on honeybees of low-level exposure to these chemicals. Twelve months on, yet-to-be-published research from federal labs suggests that at very low doses they can foster the growth of bacterial pathogens such as nosema.

In Europe the noose is tightening around the pesticide companies. Following the death of two-thirds of Germany's bees along the Rhine in summer 2008 after what Bayer said was a misapplication of clothianidin, the German Office for Consumer Protection and Food Safety suspended the approval of eight pesticides. Ten months later, the ban is still in place for sweetcorn. Slovenia followed suit, and in Italy, where up to 40% of the country's one million hives were esti-mated to have been wiped out in 2007-8, the government issued an immediate one-year suspension in September 2008 on clothianidin, imidacloprid, fipronil and thiamethoxam, used on oilseed rape, sunflowers and sweetcorn.

But the real breakthrough came at the beginning of 2009, when tough European Union regulations to ban bee-toxic chemicals, first proposed by Green MEPs, became law. Under the new EU rules, pesticides suspected of having harmful effects on honeybee colonies – which according to

the chemical industry means up to a fifth of the most important substances on the market – will be phased out and new
chemicals approved only if it can be demonstrated that there
are "no unacceptable acute or chronic effects on colony
survival and development, taking into account effects on
honeybee larvae and behaviour".

Despite the British government's resistance to the
Europe-wide ban, one of the UK's leading supermarkets,
the Co-op, announced shortly afterwards that it would
prohibit suppliers of its own-brand fruit and vegetables from
using eight pesticides until they are shown to be safe to bees.

Pesticide regulation is also coming under closer scrutiny
in the States. In August 2008, the Natural Resources
Defense Council filed a lawsuit accusing the Environmental
Protection Agency (EPA) of withholding data that the
NRDC believes implicate clothianidin in bee deaths. And
California could become the first state to ban bee-toxic
chemicals after it initiated a reevaluation of 282 pesticide
products. This followed the discovery of imidacloprid levels
in ornamental plants 20 times higher than the lethal dose
for honeybees.

Peter Neumann, the Swiss bee pathologist heading up
Coloss, the international network investigating honeybee
losses, is disturbed by gaps in our knowledge of the potential
sublethal consequences of pesticides. He says in particular:
"We need to look for possible effects on reduced life

expectancy of winter bees." If they die prematurely, the life-cycle of the colony is disrupted and it withers and dies.

Tests being developed by the USDA's Pettis and the EPA to measure sublethal effects of pesticide exposure on larvae development in bees suggest that the US regulator has finally woken up to the fact that current toxicity tests on adult bees are inadequate.

A year after Coloss received EU funding to collect and standardise bee loss figures across 35 countries and to piece together information that might identify the killer, Neumann holds up his hands in defeat. "We don't know what exactly is killing the bees," he says. But he maintains that the ubiqui-tous varroa mite is the main protagonist, with pesticides, pathogens and poor nutrition – particularly the pollen in sunflowers and blueberries – playing supporting roles.

It is too early to say how bees fared through the winter of 2008-9, but initial indications from California's almond orchards suggest that survival rates are better this year because beekeepers doubled the protein supplements they fed their apian workers in the autumn.

However, reports that the threat of extinction is subsid-ing are premature. Speaking to us from Central Valley in the middle of March 2009, Hackenberg says he may have lost another two thirds of his colonies for the third year running. "Those that didn't chase [pollinate] pumpkins and apples are looking good, but the others …"

With almond growers forced to cut their acreage in 2009 because of severe water shortages, prices for hive rentals could drop by as much as a third in 2010. If that happens, Hackenberg warns: "Beekeepers won't be able to afford to feed their bees all the supplements they have this year, so then what will happen? We're not out of the woods yet."

On a positive note, Bayer's attitude has changed. Although it still says there is no proof that its pesticides cause CCD, its scientists are now sitting down with US beekeeping representatives to address some of their concerns, including inadequate toxicity tests. "Last year they wouldn't return my phone calls; now it's a new ball game," says Hackenberg.

Although the furore over bee deaths may ultimately mean curtains for neonicotinoids such as imidacloprid and clothianidin, a new class of pesticide was registered in the US in 2008 that is so sophisticated Bayer claims "it distributes its active ingredient, spirotetramat, upwards and downwards in the plant – in a revolutionary 'two-way systemicity' – to find and eliminate even hidden pests". So maybe the next round is just beginning.

Awareness of the essential role that honeybees play in providing the food on our dinner plates, and the crisis they are facing, has grown so much in the UK since the first edition of this book that British companies such as Rowse

Honey and the cereal-maker Jordans, as well as the children's TV programme Blue Peter, are falling over themselves to donate to bee research or give away bee-friendly plants and seeds. Against this groundswell of concern and higher bee losses in 2008, the British government stumped up an additional £2m for a new research programme on pollinators – far short of the £8m demanded by the British Beekeepers' Association, but a start.

People want to know how they can help. "What if we all had a hive in the backyard?" was a question we asked last year. Since then, the BBKA has heard from around 3,000 prospective new keepers. But parliament's spending watchdog, in a report in March 2009, claimed that amateur beekeepers were partly to blame for almost one in three of the dead hives reported by BBKA members in 2008 because they didn't know how to spot and combat disease. While we would agree that the education of beekeepers is paramount for controlling varroa and nosema, cutbacks in funding of research into honeybee pests and pathogens in the UK have deprived beekeepers of the knowledge they require to fight the insects' foes. One of the aims of Coloss is to produce standardised treatment and education across Europe for controlling the viral vector.

Planting bee-friendly gardens throughout our expanding towns and cities is another nice idea. If we all threw away the weedkiller, ditched the decking and stopped concreting over our front lawns, we'd be helping numerous insects and birds. But that alone is not enough.

Nothing less than a transformation in our treatment of honeybees will avert disaster. We are the ones killing the honeybee through ignorance, unsustainable agricultural practices and dangerous use of chemicals. Urgent change is needed. But can we achieve it?

CCD TIMELINE

November 2006
- Migratory beekeeper Dave Hackenberg discovers empty hives in Florida

December 2006
- Penn State University and Pennsylvania department of agriculture issue the first report on the phenomenon initially known as fall dwindle disease

January 2007
- The disease is renamed colony collapse disorder and a working team is set up, comprising academics and private and federal scientists
- In the UK, John Chapple, chair of the London Beekeepers Association, loses most of his bees to what he dubs the "Marie Celeste syndrome"

March 2007

- Congressional hearing into CCD
- An international network is launched at a meeting of bee experts in the Netherlands. Swiss bee pathologist Peter Neumann coordinates the Coloss (prevention of honeybee colony losses) group

April 2007

- A formal CCD working team is established, chaired by Diana Cox-Foster of Penn State University and Jeff Pettis of the USDA agricultural research service
- A two-day workshop on CCD is held at the federal bee lab at Beltsville, Maryland
- Taiwan reports that 10 million honeybees have disappeared
- Canada suffers 30% honeybee mortality – twice the normal winter rate – but the government attributes it to cold spring weather
- Germany, Spain, Portugal, Italy, Poland, Austria, Belgium, the Netherlands, Slovenia and Croatia all report severe honeybee losses

June 2007

- The CCD working team issues an action plan
- A survey sponsored by Bee Alert Technology identifies CCD in 35 US states and Puerto Rico

- Green MEP Hiltrud Breyer tables an emergency motion in the European Parliament calling for systemic neonicotinoid pesticides to be banned in Europe while their possible role in honeybee deaths is investigated

July 2007

- Almost a quarter of US beekeeping operations have been struck by CCD, an Apiary Inspectors of America survey suggests. The USDA later confirms that CCD has claimed a third of US honeybee colonies

August 2007

- The US Farm Bill promises $100m for bee research and conservation

September 2007

- A metagenomic survey led by Cox-Foster and W Ian Lipkin, an infectious disease expert from Columbia University, demonstrates a strong correlation between Israeli acute paralysis virus and CCD
- The Australian Honeybee Industry Council quickly defends Australian bees against charges of carrying the virus to the US
- Spanish scientists believe the CCD killer is the parasite *Nosema ceranae*

October 2007

- The Nature documentary Silence of the Bees is screened on PBS
- CBS's 60 Minutes examines CCD
- The European Parliament supports at the first reading a new regulation on the placing of pesticides on the market, which could lead to substances that are toxic to bees not being approved

November 2007

- UK farming minister Lord Rooker tells parliament the honeybee population in Britain could be wiped out in 10 years
- French agriculture minister Michel Barnier tells his parliament that bee deaths have been confirmed in farming areas where products often claimed to be responsible, such as Regent and Gaucho, had not been used

December 2007

- Emergency session on CCD at the Entomology Society of America's AGM

January 2007

- Hackenberg loses two thirds of his hives
- Maryann Frazier of Penn State University tells the

American Beekeeping Federation conference that 43 pesticides have been identified in 92 pollen samples, 17 of them in one piece. Chemicals used in hives to control varroa mites were found at high levels

February 2008

- Adee Honey Farms, the US's largest commercial bee operator, reveals it has lost 40% of its 70,000 hives
- Pettis is made co-ordinator of five-year federal programme to improve honeybee health, survivorship and pollination availability
- Häagen-Dazs provides $250,000 for research into bee health at Penn State University and the University of California, Davis
- 1.2 million honeybee hives arrive in California's Central Valley. The almond pollination goes ahead without problems, despite the second year of CCD
- French food safety agency Afssa unveils the results of a long-term study of honeybee mortality that finds no evidence of a statistical relationship between the pesticide imidacloprid (Gaucho) and colony deaths

March 2008

- As the Farm Bill stalls, a bipartisan group of 18 senators, including Hillary Clinton, reiterates calls for $20m to fund bee research

- In France, reported bee mortality rates shoot up to 60%

April 2008

- An Apiary Inspectors of America (AIA) survey reveals that 36% of colonies died over the 2007-8 winter – a 14% increase on the previous year. More than a third of beekeepers across 29 states attributed some of the deaths to CCD-like symptoms
- Coloss meets in Greece. It now has 28 members, including the US, China and South Korea
- The UK government launches its honeybee health strategy for consultation, amid calls from the British Beekeeping Association for £8m for bee health research over five years
- The department for the environment, food and rural affairs reports that one in four bee colonies visited by its inspectors in England and Wales has died over the winter

May 2008

- The AIA issues an Israeli acute paralysis virus fact sheet telling beekeepers: "Concern over IAPV detection is not warranted." It is not the CCD killer

SOURCES

Introduction

Agnew, Singeli. *The Almond and the Bee*, San Francisco Chronicle, October 14 2007

Almanac, Almond Board of California, 2007

Shultz, Doug, producer. *Silence of the Bees*, Nature, PBS thirteen WNET New York/Partisan Pictures, 2007

Traynor, Joe. *Almond Pollination Handbook*, Kovak Books, 1993

vanEngelsdorp, Dennis; Hayes, Jerry; Pettis, Jeff. *Preliminary results concerning the loss of honey bee colonies over the winter 2007-08*, survey conducted by the Apiary Inspectors of America and the USDA Beltsville Bee Research Lab, April 2008

Chapter 1

Allsop, Karen and Brand Miller, Janette. *Honey revisited: a reappraisal of honey in pre-industrial diets*, British Journal of Nutrition 75, 1996

Ellis, Hattie. *Sweetness & Light: The Mysterious History of the Honeybee*, Sceptre 2005

Genome Research, October 2006. www.genome.org

Horn, Tammy. *Bees in America*, University Press of Kentucky, 2005

Kellar, Brenda. *Honey Bees Across America*. Paper presented at Oregon State Beekeepers Association 2004 conference, www.orsba.org

Lemanski, Jay S. *The Rectitudines Singularum Personarum: Anglo-Saxon Landscapes in Transition*. August 2005

Out of Africa: Scientists Uncover History of Honey Bee. ScienceDaily, October 2006. www.sciencedaily.com

Tennessee Beekeepers Association. *History of Honey Bees*. www.tnbeekeepers.org

The Honeybee Genome Sequencing Consortium. *Insights into social insects from the genome of the honeybee, Apis Mellifera*. Nature, vol 443, October 26 2006

Virgil. *Georgics*, Penguin Classic, 1982

Wilson, Bee. *The Hive*, John Murray, 2004

Chapter 2

Buchmann, Stephen and Nabhan, Gary. *The Forgotten Pollinators*, Island Press, 1997

Darwin, Charles. *On the Origin of Species*, John Murray, 1859

Michener, Charles D. *The Bees of the World,* Johns Hopkins University Press, 2000

von Frisch, Karl. *The Dance Language and Orientation of Bees,* The Belknap Press of Harvard University, 1967

Winston, Mark L. *The Biology of the Honey Bee*, Harvard University Press, 1987

Chapter 3

Anklagemyndighed v Ditlev Bluhme, judgment of European Courts of Justice, December 1998

Chapman, Nadine; Lim, Julianne; Oldroyd, Benjamin. *Population genetics of commercial and feral honey bees in Western Australia,* Journal of Economic Entomology 101 (2) 2008

Fries, Ingemar; Imdorf, Anton; Rosenkranz, Peter. *Survival of mite infested (Varroa destructor) honey bee (Apis mellifera) colonies in a Nordic climate,* Apidologie 37, 2006

Fries, Ingemar; Bommarco, Riccardo. *Possible host-parasite adaptations in honey bees infested by Varroa destructor mites,* Apidologie 38, 2007

Groehn, Ulf and Peterson, Hjalmar. *Læsø bees – last survivors of a pure northern species,* Bee Improvement and Bee Breeders Association, 2000

Jones, Julia C; Myerscough, Mary R; Graham, Sonia; Oldroyd, Benjamin. *Honey Bee Nest Thermoregulation: Diversity Promotes Stability,* Science, volume 305, July 16 2004

Kryer, Per. *The dark bee on Læsø,* email, 2008

Lodesani, Marco and Costa, Cecilia. *Beekeeping and conserving biodiversity of honeybees,* Northern Bee Books, 2005

Mattila, Heather and Seeley, Thomas. *Genetic diversity may be the key to new honey bee colony success,* National Research Initiative Competitive Grant Program No 11, 2007

Oldroyd, Benjamin. *What's killing America's Honey Bees?,* PLoS Biology, Volume 5, issue 6, June 2007

Schiff, NM and Sheppard, WS. *Genetic analysis of commercial honey bees (Hymenoptera: Apidae) from the Southeastern United States,* Journal of Economic Entomology 88 (5) 1216-1220, 1995

Schiff, NM; Sheppard, WS; Loper, GM; Shimanuki, H. *Genetic diversity of feral honey bee (Hymenoptera: Apidae) populations in the southern United States,* Annals of the Entomological Society of America, 87 842-84, 1994

Seeley, Thomas and Tarpy, David. *Queen promiscuity lowers disease within honeybee colonies,* Proceedings of the Royal Society, 2006

Tarpy, David. *Genetic diversity within honeybee colonies prevents severe infections and promotes colony growth,* Proceedings of the Royal Society, November 19 2002

Chapter 4

Anderson, J. *"Isle of Wight disease" in bees,* Bee World 11 37-42, 1930

Aristotle. *History of Animals VII-X,* Loeb Classics Library, 1991

Bailey, L. *The "Isle of Wight disease": the origin and significance of the myth,* Bee World 45 32-37, 1964

Beuhne, R. *Bee mortality,* Journal of the Department of Agriculture of Victoria 7 149-151, 1910

Carr, EG. *An unusual disease of honey bees,* Journal of Economic Entomology 11 347-351, 1918

Faucon, JP et al. *Honey bee winter mortality in France in 1990 and 2000,* Bee World 83 14-23, 2002

Finley, JS; Camazine, S; Frazier M. *The epidemic of honey bee colony losses during the 1995-1996 season*, American Bee Journal 136 805-808, 1996

Kauffeld, NM; Everitt, JH; Taylor, EA. *Honey bee problems in the Rio Grande Valley of Texas*, American Bee Journal 116 220-222, 1976

Kulincevic, JM; Rothenbuhler, WC; Rinderer, TE. *Disappearing disease, part 1. Effects of certain protein given to honey-bee colonies in Florida*, American Bee Journal 122 189-191, 1982

Mraz, C. *Disappearing disease south of the border*, Gleanings in Bee Culture 105 198, 1977

Oertel, E. *Many bee colonies die of an unknown cause*, American Bee Jounal 105 48-49, 1965

Olley, K. *Those disappearing bees*, American Bee Journal 116 520-521, 1976

Report of the Commissioner of Agriculture for the year 1868, US Government Printing Office, Washington DC 272-281, 1869

Thurber, PF. *Disappearing – yes; disease – no!*, Gleanings in Bee Culture 104 206-261, 1976

Underwood, Robyn and vanEngelsdorp, Dennis. *Colony collapse disorder: have we seen this before?*, Bee Culture 135 (7) 13-18., 2007

Wilson, WT and Menapace, DM. *Disappearing disease of honey bees: a survey of the United States*, American Bee Journal 119, part one 118-119, part two 184-186, 1975

Witherell, PC. *Conference on the disappearing disease of honey bees*, American Bee Journal 115.300, 1975

Chapter 5

Adee, Richard. *Statement for the committee on agriculture, US House of Representatives*, March 29 2007

Anderson, Denis and East, Iain. *The Latest Buzz About Colony Collapse Disorder*, letter to Science magazine, volume 319, February 8 2008

Bee Alert Technology. *CCD survey*, June 2007. http:// beealert.blackfoot.net/~beealert/USshaded.pdf

Beekeepers seek ways to protect fragile industry from new and deadly threats, Canadian press, January 21 2008

Berenbaum, May. *Statement for the committee on agriculture, US House of Representatives*, March 29 2007. http://www7. nationalacademies.org/ocga/testimony/Colony_Collapse_ Disorder_and_Pollinator_Decline.asp

CCD Steering Committee. *Colony Collapse Disorder Action Plan*, June 20 2007. www.ars.usda.gov/is/br/ccd/ccd_actionplan.pdf

Cox-Foster, Diana. *Statement for the committee on agriculture, US House of Representatives*, March 29 2007. http://maarec. cas.psu.edu/CCDPpt/CoxFosterTestimonyFinal.pdf

Cox-Foster, Diana; Lipkin, W Ian et al. *A Metagenomic Survey of Microbes in Honey Bee Colony Collapse Disorder*, Science Express, September 16 2007. www.scienceexpress.org

Ellis, James and Munn, Pamela. *The worldwide health status of honey bees*, Bee World 86 (4) 88-101, December 2005

Frazier, Maryann. Recording of presentation at American Beekeeping Federation, National Beekeeping Conference, January 2008

Hayley, Julia. *Asian Parasite Killing Western Bees – Scientist*, Reuters, July 19 2007

Huck, Peter. *The strange case of the vanishing bee*, The Guardian, June 27 2007

Latsch, Gunther. *Are GM Crops Killing Bees?*, Der Spiegel, March 22 2007

Ritter, Wolfgang, *Bee death in the USA: Is the honey bee in danger?* Unpublished

Ritter, Wolfgang. *Severe Bee Losses also in Germany: somewhat expected and not at all mysterious*, Beekeeping journal of the German Agricultural Book-Publishing House, March 2008. http://www.beepathology.com/html/info_germany.html

Rooker, Lord. *Bees debate, House of Lords*, Hansard, column 1113, November 27 2007

Root, Amos Ives. *The ABC of Bee Culture*, Kessinger Publishing, LLC, March 30 1947

Svensson, Borje. *Silent spring in northern Europe?* www.beekeeping.com/intoxications/silent_spring.htm

Taiwan stung by millions of missing bees, Reuters, April 25 2007

vanEngelsdorp, D; Cox-Foster, D; Frazier, M; Ostiguy, N; Hayes, J. *Fall-Dwindle Disease: Investigations into the causes of sudden and alarming colony losses experienced by beekeepers in the fall of 2006*, Mid-Atlantic Apiculture Research and Extension Consortium, December 15 2006. http://maarec.cas.psu.edu/pressReleases/FallDwindleUpdate0107.pdf

Chapter 6

Aubert, M; Faucon, J-P; Chauzat, M-P. *Prospective multifactorial study: Effect of microbial and parasitic agents and pesticide residues on the evolution of domestic bee colonies under natural conditions,* Afssa study, February 2008

BayerCropScience. *BayerCropScience is convinced of the bee safety of Gaucho,* March 2004. www.bayercropscience.com

Bonmatin, JM et al. *Imidacloprid in soils, in plants and in pollens,* Analytical Chemistry, May 2003

Bortolotti, L et al. *Effects of sub-lethal imidacloprid doses on the homing rate and foraging activity of honey bees,* Bulletin of Insectology 56 (1) 63-67, 2003

Browner, Carol. EPA speech, June 8 2000 www.epa.gov/history/topics/legal/o3.htm

Buffin, David. *Imidacloprid,* Pesticides News No 62, 22-23, December 2003

Carson, Rachel. *Silent Spring*, Penguin Classics, 2000

Crop Protection Association. Power Point presentation, November 2007

Cox, Caroline. *Insecticide factsheet,* Journal of Pesticide Reform, Vol 21, Spring 2001

Cummins, Joe. *Requiem for the honeybee,* Science in Society, April 24 2007. www.i-sis.org.uk/requiemforthehoneybee.php

Decortye, A; Lacassie, E; Pham-Delegue, MH. *Learning performances of honeybees are differentially affected by imidacloprid according to the season,* Pest Management Science 59 (3) 269-78, 2003

Decortye, A et al. *Imidacloprid impairs memory and brain metabolism in the honeybee*, Pesticide Biochemistry and Physiology 78 (2004) 83-92, 2004

Fishel, Frederick. *Pesticide toxicity profile: Neonincotinoid pesticides*, University of Florida, October 2005 http://edis.ifas.ufl.edu

Frazier, Maryann. Recording of presentation at American Beekeeping Federation, National Beekeeping Conference, January 2008

French beekeepers abuzz with worry over dying bees, Associated French Press, January 1 2008

Governmental report claims Bayer pesticide Gaucho responsible for bee deaths. Press release, Coalition against BAYER-dangers, November 25 2003. www.cbgnetwork.org

Hackenberg, Dave. *Letter to growers*, March 14 2007. www.thedailygreen.com/environmental-news/blogs/bees/

Laurent, FM; Rathahoa, E. *Distribution of imidacloprid in sunflowers following seed treatment,* Journal of Agricultural Food Chemistry, December 2003

Mares, Bill. *Bees Besieged: One Beekeeper's Bittersweet Journey to Understanding*, AI Root Company, 2005

Medrzycki, P et al. *Effects of imidacloprid administered in sub-lethal doses on honey bee behaviour: lab tests*, Bulletin of Insectology 56 (1) 59-62, 2003

Rose, Robyn; Dively, Galen; Pettis J. *Effects of Bt corn pollen on honey bees: emphasis on protocol development*, Apidologie 38 368-377, 2007

Schmuck, R. *No causal relationship between Gaucho and French bee syndrome,* 1999. www.bayercropscience.fr/upload/flb/88/ nocausal_relationship.pdf

The co-ordination of French Beekeepers. Stricken sector of production in apiculture: Beekeeping hit in full flight, translated by Peter Dillon, October 2000. www.beekeeping.com/articles/us/ gaucho/gaucho_france.htm

Tukey, Paul. *What's killing the bees?,* Environmental News Network, November 2007

Chapter 7

Carreck, Norman. *CCD: A View From Across the Pond,* Bee Culture, January 2008

Higes, Mariano. Recording of presentation at American Beekeeping Federation National Beekeeping Conference, January 2008

Managing Varroa, Department for Environment, Food and Rural Affairs, 2005

Martin, Stephen J. *The role of varroa and viral pathogens in the collapse of honeybee colonies: a modelling approach,* Journal of Applied Ecology. (38) 1082-1093, 2001

Martin, Stephen J. *Acaricide (pyrethroid) resistance in Varroa destructor.* Bee World 85 (4) 67-69, December 2004

Martin, Stephen J (editor). *Apiculture research on varroa; original research on bees in the 21st century.* International Bee Research Association, 2007

Oldroyd, Benjamin P. *Coevolution while you wait: Varroa jacobsoni, a new parasite of western honeybee,* Tree, volume 14, number 8, August 1999

Sanford, Malcolm T. *Two decades of varroa,* Beekeepers Quarterly, February 2008.

Sumpter, D and Martin, SJ. *The dynamics of virus epidemics in varroa-infested honey bee colonies,* Journal of Applied Ecology. (73) 51-63, 2004

Suszkiw, Jan. *Russian Bee Earning its Stripes,* Agricultural Research magazine, October 2001. www.ars.usda.gov/is/AR/archive/oct01/bee1001.htm

Thompson, Helen et al. *First report of Varroa destructor resistance to pyrethroids in the UK,* Apidologie 33 357-366, 2002

Traynor, Joe. *CCD and the varroa-virus complex,* Beekeeper Newsletter, March 2008

Vejsnaes, Flemming. *Varroa Treatments, Denmark: Questionnaire 2005,* Beekeepers Quarterly, December 2007

Chapter 8

American Farmland Trust. *What's happening to our farmland?* www.farmland.org, 2002

Braunstein, Martin. *Chronicle of a crisis announced,* 2008

Hornets hit France and could reach Britain, The Telegraph, February 22 2007

Jha, Alok. *Pollution stifling flowers' scents,* The Guardian, April 14 2008

Lean, Geoffrey and Shawcross, Harriet. *Are mobile phones wiping out our bees?,* The Independent, April 15 2007

Lindsey, Rebecca. *Buzzing about climate change,* Earth

Observatory, September 7 2007. http://earthobservatory.nasa. gov/Study/Bees/

Mussen fingers prime suspects in BSI: the case of the disappearing bees, University of California News and Information Outreach, October 18 2007. http://news.ucanr.org/newsstorymain. cfm?story=1040

Chapter 9

Agnew, Singeli. *The Almond and the Bee,* San Francisco Chronicle, October 14 2007

Almanac, Almond Board of California, 2007

Mares, Bill. *Bees Besieged: One Beekeeper's Bittersweet Journey to Understanding,* AI Root Company, 2005

Nordhaus, Hannah. *The Silence of the Bees,* High Country News, March 19 2007

Traynor, Joe. *Almond Pollination Handbook,* Kovak Books, 1993

Chapter 10

Buchmann, Stephen and Nabhan, Gary. *Forgotten Pollinators,* Island Press, 1997

Carson, Rachel. *Silent Spring,* Penguin Classics, 2000

Committee on the Status of Pollinators in North America. *Status of Pollinators in North America,* National Academy of Sciences, 2006

Cronon, William. *Changes in the Land: Indians, Colonists and the Ecology of New England,* Hill and Wang, 1983

Diamond, Jared. *Collapse: How societies choose to fail or succeed*, Viking, 2004

Flottum, Kim, *Does the USDA Even Care About the Bee Crisis?*, Daily Green, April 2008

Greenleaf, S and Kremen, C. *Wild bees enhance honey bees' pollination of hybrid sunflower*, Proceedings of the National Academy of Sciences, August 29 2006

International Assessment of Agricultural Science & Technology. *Agriculture – the need to change report*, April 15 2008. www.agassessment.org

Leidig, Michael. *Honeybees in US facing extinction*, The Telegraph, March 14 2007

Levek, Amy. *The Xerces Society/The Buzz on Bees*, The Watch, August 2 2007

Lochhead, Carolyn. *Farm bill complicates plight of honeybees*, San Francisco Chronicle, April 19 2008

Protecting the pollinators, Food and Agricultural Organisation of the UN, 2005

Shultz, Doug, producer. *Silence of the Bees*, Nature, PBS thirteen WNET New York/Partisan Pictures, 2007

Weisman, Alan. *The World Without Us*, Virgin Books, 2008

Wilson, EO. *The Future of Life*, Abacus, 2003

Wilson, EO. *The Creation*, WW Norton & Company, 2006

Chapter 11

Apiary Inspectors of America. *IAPV Fact Sheet*, May 2008

INDEX

ACKNOWLEDGMENTS

Our thanks go to all the people who gave their time to be interviewed for this book, or who responded to our emails, and to those we visited, those who generously put us in touch with their contacts, and those who provided us with new information on this ongoing story.

In particular we would like to thank: Joe Traynor for his generous time and support in California, and his hive inspector, Bill Mathewson, who showed us around and introduced us to many migratory beekeepers; Dave Hackenberg for sharing his stories and theories with us; Bret Adee for taking us out to see his thousands of empty hives; Diana Cox-Foster for inviting us to visit her at Pennsylvania State University, and for her patience, cooperation and candour in explaining and showing us the experiments she is conducting; Dennis vanEngelsdorp for answering our endless questions and helping us with new lines of inquiry; Kim Flottum, editor of Bee Culture magazine, whose regular blog on The Daily Green kept us up to date and entertained; Wolfgang Ritter for the European perspective; and Norman Carreck, whose knowledge proved a valuable starting point for this project.

Writing this book in such a short timescale would have been impossible without Paul Feldman's diligent research and Samantha Rennie, Richard Courtin and Cecilia Giussani's translations.

Thanks also to Guardian journalists John Vidal, for his infectious enthusiasm, and Felicity Lawrence, for her reading suggestions.

This book would not have happened without Lisa Darnell at Guardian Books asking Alison to write it and having faith in us to deliver it, and our editor, Phil Daoust.

We'd also like to thank the science librarians at the British Library – the most wonderful place to work in London – for their help. And finally, our friends and family for their support and encouragement.